Geodemographics, GIS and Neighbourhood Targeting

Mastering GIS: Technology, Applications and Management series

Geodemographics, GIS and Neighbourhood Targeting

Richard Harris
University of Bristol

Peter Sleight
Target Market Consultancy

and

Richard Webber
Visiting Professor, Centre for Advance Spatial Analysis, UCL

John Wiley & Sons, Ltd

Other Wiley Editorial Offices

John Wiley & Sons Inc., 111 River Street, Hoboken, NJ 07030, USA

Jossey-Bass, 989 Market Street, San Francisco, CA 94103-1741, USA

Wiley-VCH Verlag GmbH, Boschstr. 12, D-69469 Weinheim, Germany

John Wiley & Sons Australia Ltd, 33 Park Road, Milton, Queensland 4064, Australia

John Wiley & Sons (Asia) Pte Ltd, 2 Clementi Loop #02-01, Jin Xing Distripark, Singapore 129809

John Wiley & Sons Canada Ltd, 22 Worcester Road, Etobicoke, Ontario, Canada M9W 1L1

Wiley also publishes its books in a variety of electronic formats. Some content that appears in print may not be
available in electronic books.

Library of Congress Cataloging in Publication Data

Harris, Richard, 1973–
     Geodemographics, GIS, and neighbourhood targeting / Richard Harris, Peter Sleight, Richard Webber.
      p.   cm.
     Includes bibliographical references.
     ISBN 0-470-86413-3 (cased) — ISBN 0-470-86414-1 (pbk.)
     1. Target marketing.   2. Geographic information systems.   3. Demography.
     I. Sleight, Peter, 1944–   II. Webber, Richard, 1947–   III. Title.
     HF5415.127.H37 2005
     304.6′0285—dc22

                                                     2004021912

British Library Cataloguing in Publication Data

A catalogue record for this book is available from the British Library

ISBN-13: 978-0-470-86413-5 (H/B)
ISBN-13: 978-0-470-86414-2 (P/B)

Typeset in 11/13 pt Times by Integra Software Services Pvt. Ltd, Pondicherry, India
Printed and bound in Great Britain by Antony Rowe Ltd, Chippenham, Wiltshire
This book is printed on acid-free paper responsibly manufactured from sustainable forestry in which at least two
trees are planted for each one used for paper production.

For Rhys

Contents

Contents

Contents

List of Case Study Contributors

Stewart Berry
Caliper Corporation
Newton, MA
USA

Martin Callingham
School of Geography
Birkbeck College
University of London
UK

Keith Dugmore
Demographic Decisions Ltd
London
UK

Gordon Farquharson
Streetwise Analytics Ltd
Windsor, Berkshire
UK

Peter Furness
Peter Furness Ltd, Surrey
UK

Barry Leventhal
Teradata, a division of NCR
UK

List of Case Study Contributors

David Miller
Claritas, Arlington, VA
USA

Scott Orford
Department of City and Regional Planning
Cardiff University
UK

Tom Williamson
Institute of Criminal Justice Studies
University of Portsmouth
UK

Preface

As 'lead author' my colleagues have invited me to write the preface and acknowledgements for this book. This may be a strategy to distance themselves from any blame. However, I suspect it is more an act of courtesy and one that I appreciate. Still, I regard the production of this book as very much a team effort and I am extremely grateful to Peter and to Richard for all that they have contributed to it. In other words, you guys aren't getting off that easily!

The publication of this book reflects a chance to put into print our longstanding interest and involvements in – as the title identifies – geodemographics, GIS and neighbourhood targeting. Together with our contributors we have spent a rather frightening amount of time preoccupied by these fields of employment and research – over 100 years in fact! Had those years not been concurrent we might have had a chance to have met some of the originators of present-day geodemographics – Charles Booth in London and the 'Chicago School' of urban sociologists working in . . . well, the name is a giveaway!

How much of the consumer behaviours, preferences and socioeconomic characteristics of individuals are revealed by where they live? The answer to that question is of enduring interest to us and so, therefore, are recently developed methods of spatial analysis that can be applied to new sources of lifestyle data to search for and quantify the elusive neighbourhood effects which are said to lie at the heart of geodemographics. Do birds of a feather flock together? Can we really know something about who you are if we know where you live? These are questions that interest us enormously and, although we don't have all the answers, we do offer some suggestions as to why it is that geodemographics 'works', has grown and has evolved as an industry and method of analysis, and has been applied successfully in a variety of contexts.

The purpose of this book is to consider the relevance, strengths and limitations of the geodemographic idea for public- and private-sector decision making. Its authorship is somewhat unusual in that we approach the subject from differing perspectives – both academic and commercial.

Appealing to both audiences is by no means an easy task but, we hope, a worthwhile aim. Certainly, we have found our shared discussions useful and illuminating. Those at the front line of marketing and commercial decision making need to get things done with the most effective tools and the most accurate data that currently are available. Those in 'ivory towers' (or, more likely, concrete carbuncles) often have the luxury of time to critique (albeit less time than they used to) and don't always understand business environments, the nature of business relationships or the strategic needs and requirements of the users of geodemographic systems. At the same time, academic writers have offered a number of extremely useful insights concerning geodemographics, its strengths, weaknesses and possible development, but these are not always brought to the attention of users because the 'two sides' don't talk as often as they might. In this book we look at best practice when using geodemographic methods, software and systems, and so aim to balance academic theorizing with the practicalities of commercial life.

Because the seeds of this book lie in an MSc completed during my first stint at Bristol and germinated thereafter with my jaunts eastwards, then westwards along the Great Western Railway, there are a number of people to whom I owe thanks. It's not much by way of recompense but I am grateful to the following for providing support, advice and, most importantly, friendship over that time: Siân, mum, dad, Nikki, Wilma, Margaret, Elwyn, Ann, Bill, Ed, Revd. Andy, Jo, Liz, Scott, Katy, Kate, Fiona, Paul Longley, Victor Mesev, colleagues at Birkbeck College (especially Andrew Jones, Martin Frost, Zunqiu Chen, Melanie Roy, Tessa Hilder, John Shepherd and Dave Unwin), the GIS group at the University of Glamorgan (Gary Higgs, Chris Brunsdon, George Taylor, Mitch Langford, Mark Ware and Dave Kidner), the spatial modelling group at the University of Bristol (Ron Johnston, Paul Plummer, Les Hepple, Tony Hoare, Kelvyn Jones) and the members of various house groups at Christ Churches Downend and Clifton. To all and any omissions: 2 Corinthians 9:15.

All three authors would like to thank our families, the contributors to this book – Martin Callingham, Keith Dugmore, Scott Orford, Dave Miller, Stewart Berry, Barry Leventhal, Peter Furness, Gordon Farquharson, Tom Williamson and Gillian Harper – and the staff at Wiley, especially Keily Larkins and Lyn Roberts. Finally, we are grateful to the following companies for supplying data and software that have been incorporated during the longer course of this project: Experian, Claritas UK, Caliper Corporation and CACI UK. Any errors in interpretation or analysis are our own.

Rich Harris
Bristol

1
Introducing Geodemographics

Learning Objectives

In this chapter we will:

- Define geodemographics as the analysis of people by where they live.

- Explore why it is a useful framework for public- and private-sector decision making.

- Offer initial explanations and worked examples of how geodemographics 'works'.

- Introduce Tobler's 'first law of geography' and the concept of spatial autocorrelation.

- Present two of the specially commissioned case studies that appear throughout this book. The first is authored by Martin Callingham, formerly the group manager of market research for Whitbread plc and is about using the geodemographic approach to model price sensitivity in the restaurant market. The second is by Keith Dugmore of Demographic Decisions Ltd and chair of the Demographics User Group, and is about using geodemographics in the public sector.

Geodemographics, GIS and Neighbourhood Targeting   Richard Harris, Peter Sleight and Richard Webber
© 2005 John Wiley & Sons, Ltd   ISBNs: 0-470-86413-3 (HB); 0-470-86414-1 (PB)

Introduction

Geodemographics is the 'analysis of people by where they live' (Sleight, 1997, p. 16). It is the suggestion that *where* you are, says something about *who* you are; that knowing where someone lives provides useful information about how that person lives. To quote some product advertising, it is the possibility that 'we know who you are, because we know where you live.' Figure 1.1 illustrates this link between people and places. It is a simple idea – one that has shown itself to be of commercial value and the catalyst of a rapidly growing and globalizing industry.

The purpose of this book is to consider the relevance, strengths and limitations of the geodemographic idea for public- and private-sector decision making. We provide an introduction to and overview of the methods, theory and classification techniques that provide the foundations of neighbourhood analysis and commercial geodemographic products. We give examples of using geodemographic analysis effectively to target resources and offer guidance to best practice that draws upon our contributors' experiences of working within the geodemographic industry.

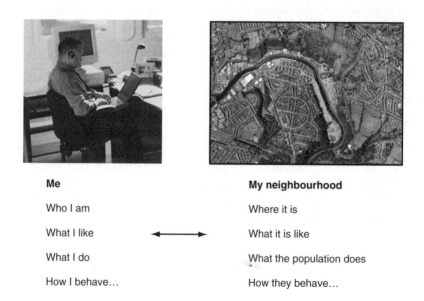

Figure 1.1 Geodemographics is 'the analysis of people by where they live', linking people to places

Within the book, particular focus is given to linking geodemographics with the theories and products of geographical information science, notably geographic information systems (GIS). Our aim throughout is to provide reader-friendly theory and moderate statistical explanation, supported by relevant case studies, short vignettes and applied 'how to' sections that will appeal to an international and professional audience at work in business and service planning, and to students of marketing, geography or other spatial, social science. Along the way we include some formulae and mathematical notation as often these provide the most succinct and accurate way of describing a particular calculation or procedure. However, we know such notation will not suit everybody so we also take time to explain, in words, what the symbols summarize. In this chapter we introduce the core principles and ideas that underpin the use of geodemographic information for neighbourhood profiling and set out the agenda for the rest of the book.

1.1 The use of geodemographics

We are neither the first to take an interest in geodemographics; nor, we hope, will we to be the last! Over a decade ago Brown (1991, p. 221) commented that,

> [g]eodemographics has come into use as a shorthand label for both the development and the application of area typologies [neighbourhood classifications] that have proved to be powerful discriminators of consumer behaviours and aids to 'market analysis'.

The 'proof' is found in the increased value of the geodemographic market. In Britain this was estimated at a value of £25 million in 1992 (Sleight, 1997, p. 15, citing Mitchell, 1992). By 1995 the same market was valued at £54 million. In 1998, *Directions Magazine* (www.directionsmag.com) reported its 'conservative estimate' of 20,000 companies in the USA and Canada using commercial neighbourhood classifications as part of their marketing information. Weiss (2000) reports that US marketers spend an estimated $300 million annually on clustering techniques (see Chapter 7), profiling the behaviour of the nation's 100 million households: 'cluster-based marketing has gone mainstream and is now used by corporate, nonprofit, and political groups alike to target their audiences' (p. 4). As we shall show in Chapters 3, 7 and 9, the market has continued to evolve, the most recent stimulants being the release of twenty-first-century census data and the emergence of extensive 'data warehouses' associated with a growing trade in consumer-oriented data.

It is also over 10 years ago that Leventhal (1993, p. 223) recognized the potential of geodemographics to inform strategic marketing, planning and communications, presenting examples of its application to the Market Research Society under three main headings:

- **Survey design** ('samples may be stratified or selected using geodemographics, and many large-scale surveys take advantage of this facility').
- **Retail planning** ('knowledge of the types of people living in a catchment area can be a key ingredient in understanding store performance and the same information can help in deciding on store location').
- **Direct marketing** ('the selection of prospects [prospective customers] can be improved by using geodemographics, whether for direct mail, "door-to-door" distribution or sales calls').

Those three themes correspond to the market research, market analysis and direct marketing streams identified by Curry (1993, p. 200) and summarized by Figure 1.2. Curry also outlines a fourth role for geodemographics in advertising and media analysis (as do Sleight, 1997, pp. 117–21 and Webber, 1985). Each author recognizes the value of geodemographics as an important business tool that offers a type of analysis which is both understandable and operational within an applied, decision-making environment.

Figure 1.2 offers a marketing perspective on geodemographic applications. We shall also consider the role of neighbourhood analysis in public-sector policy and planning, avoiding the impression that geodemographics is a solely commercial affair, driven by the needs of market analysis and consumer profiling. Although these commercial elements certainly are important aspects of the field, Chapter 2 reveals that the origins and evolution of neighbourhood classification have an academic pedigree in urban geography, urban sociology and in urban planning. Batey and Brown (1995, p. 78) record that,

> [t]he pragmatic approach of the marketing analyst has much in common with that of the urban planner. In both cases an area classification system is required to provide up-to-date information that is actionable, and the test of a good system is whether it works in practice. It is perhaps not surprising, therefore, that classifications generated for use in public policy-making ultimately found their way into the private sector.

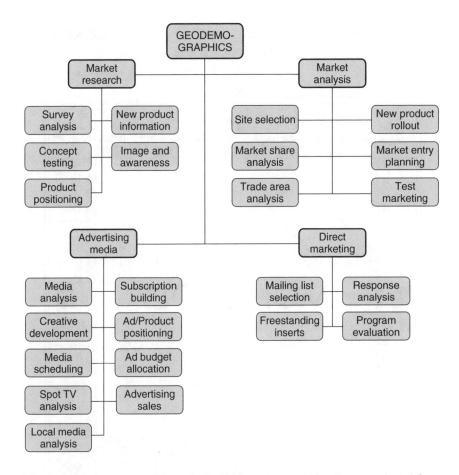

Figure 1.2 Some commercial applications of geodemographics. Source: adapted from Curry (1993, p. 200)

1.2 A simple application: opening a coffee shop in Atlantic City

A traditional, American Monopoly board (Figure 1.3) has 40 locations to land on, of which 22 are streets in Atlantic City. These streets are grouped together by eight distinct colour bands that indicate the cost of acquisition, development and rental value of the land (in the early twentieth century). Each street belongs to one and only one of the eight colour groups. Therefore, Mediterranean Avenue is either in the brown group or it isn't (it is); Boardwalk is either in the green group or it isn't (it is not, it is in the dark blue group). No street is only part in a group – each is either

5

Figure 1.3 An early geodemographic classification. Source: US Patent 2,026,082 (United States Patent and Trademark Office)

entirely in else it is entirely out (this is known as Boolean logic where a proposition is either true or false). Furthermore, if a location does not belong to the one group then it must belong to another. If necessary, locations can be placed in a ninth, residual category of 'Other' (see Table 1.1).

Table 1.1 Segmentation of sample data by street and by neighbourhood

Group	Code	Description	Streets	Outlet here?	Customers per street	% Sample per group
Brown	A	Low rental value, construction cost and low maintenance	Mediterranean Ave.	No	10	1.34
			Baltic Ave.	No	0	
Light blue	B		Oriental Ave.	No	0	4.30
			Vermont Ave.	No	13	
			Connecticut Ave.	Yes	19	
			St. Charles Place	Yes	81	
Pink	C		States Ave.	Yes	94	23.52
			Virginia Ave.	No	0	
			St. James Place	Yes	136	
Orange	D		Tennessee Ave.	No	0	18.28
			New York Ave.	No	0	
			Kentucky Ave.	Yes	124	
Red	E		Indiana Ave.	Yes	153	37.23
			Illinois Ave.	No	0	
			Atlantic Ave.	Yes	58	
Yellow	F		Ventnor Ave.	No	0	9.41
			Marvin Gardens	No	12	
			Pacific Ave.	No	0	
Green	G		North Carolina Ave.	No	34	4.57
			Penn Ave.	No	0	
Dark blue	H	High rental value, construction cost and high maintenance	Park Place	No	9	1.21
			Boardwalk	No	0	
Other	Z	Other types of location, including utilities and railroad stations	Reading Railroad	No	0	
			Municipal Jail	No	0	
			Electric Company	No	1	
			Water Works	No	0	
			Penn Railroad	No	0	0.13
			Parking Lot	No	0	
			B & O Railroad	No	0	
			Shortline Railroad	No	0	
			Internal Revenue Service	No		
			Total		**744**	**100**

Travelling clockwise around the board from 'Go', the potential revenues from the properties increase but so too do the maintenance costs. In summary, the board is a simple, area-level classification of Atlantic City that uses economic value as the variable that differentiates between types of street and permits them to be grouped together on a like-with-like basis. The classification is both mutually exclusive and collectively exhaustive (Curry, 1993) – every location is classified as belonging entirely to one and only one neighbourhood type (albeit 'other'). A classification tree is shown in Figure 1.4.

Now we have our neighbourhood classification, what can we do with it? To answer the question, imagine you are the owner of *Caffeine-II-Goad*, a small coffee chain operating in Atlantic City. You have interviewed a sample of customers visiting your stores, finding out where they have travelled from to collect their coffee. That data is shown in Table 1.1, together with information about where your existing outlets are located. You find that the highest percentage of your sample is from streets of the red type (37.23%). Assuming the sample you have taken is representative of the customers you have not sampled – a critical assumption! – then, as proprietor, make the following decision: where will you open your next outlet?

Your answer:

Your reasons:

The site we have chosen is Illinois Avenue. The rationale is that this street is of the same neighbourhood type as both Kentucky and Indiana avenues. On this basis, we feel more able to attract customers to our brand of store if we open on Illinois than if we open on, say, Baltic. Of course, the neighbourhood classification cannot guarantee that we will

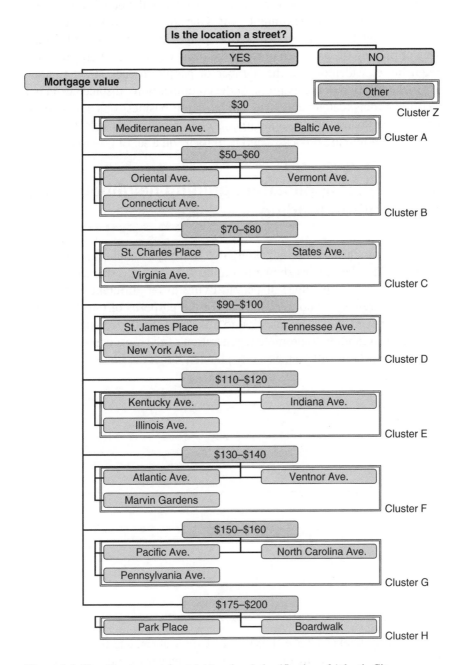

Figure 1.4 Classification tree for neighbourhood classification of Atlantic City

attract more customers but it has helped us to 'play the odds' and identify where potential customers might visit. Our strategy is not dissimilar to that of completing a colour set in Monopoly. However, it is not the only way of playing the game! In practice the decision is much more complex, requiring consideration of the number of people in each street (market size), the location of competitors' outlets (competition), distance from our existing outlets (catchments) and, not least, our business objectives (strategy). Nevertheless, the neighbourhood classification can help to inform the decision-making processes and help focus on a set of possibilities.

1.3 Another application: guiding neighbourhood regeneration funding

Imagine that instead of indicating the mortgage values of streets in Atlantic City, the monies shown in Figure 1.4 now represent the average weekly household incomes of people living in the different neighbourhoods (the values are not realistic but for the purposes of our discussion this does not matter). The question to answer now is, given that the city has received funding to promote regeneration of the most deprived neighbourhoods, where will you, as a city planner, target the funds?

Your answer:

Your reasons:

The streets we have chosen are Mediterranean and Baltic avenues. Our reasoning seems obvious – these streets appear to contain the lowest income households! What is less clear is why we chose only these two streets and did not include also, for example, Oriental, Vermont and

Connecticut Avenues. The answer is we chose the 10% most-deprived streets (the top decile, ranking the streets by average income from most to least deprived), making the assumption that deprivation correlates with average household income.

Again, in practice things are more complicated than our explanation reveals. First, the 10% threshold is an arbitrary cut-off and has the effect of segmenting the streets and consequently the resident households into two groups (or cluster types): those that are deemed deprived and those that are not; alternatively, those areas that will receive funding and those that will not. This approach could be regarded as 'heavy handed' – might those streets near the 10% threshold receive some of the funding, albeit a lower proportion?

Second, classifying 'deprived' streets is not the same thing as classifying deprived households – they exist at different geographic scales. A critical issue is how similar households are to the average for the street; how similar are they to each other? In areas of socio-economic diversity an average value (with no knowledge of the variance around that average) can give misleading results. It is possible to find pockets of deprivation in areas that otherwise appear affluent and these pockets risk being overlooked. Unfortunately we often only have the average or sum total values for an area (this is particularly true for national census data), with little knowledge of the within-area variations and population diversity (Harris and Longley, 2004).

Third, household income is not necessarily a robust indicator of deprivation. We ought to make corrections for the number of people living in the household and perhaps also allow for regional housing costs (e.g. the fact that London is more expensive to live in than Liverpool), the age of the population (retired persons would tend to have lower incomes, does this mean they are necessarily deprived?), the number of dependents (children are expensive to raise!) and so forth. Furthermore, although income may reasonably be regarded as an important factor in determining social inequalities (Hall and Pfeiffer, 2000), it is not the only measure of a person's standard of living or their level of integration within (or exclusion from) society. Deprivation indicators usually take a more rounded view, taking a selection of variables to create a multivariate profile or score of a neighbourhood's position in relation to others in the country (Lee, Murie and Gordon, 1995; ODPM, 2004). In the same way, commercial geodemographic classifications do not classify neighbourhoods based on a single variable such as mortgage value (in Figure 1.4) but take a range of data to better emphasize the differences between neighbourhood types.

1.4 Using geodemographics for retail targeting

Our example of using geodemographic classification in Atlantic City can be generalized to a retail company that takes its client list and sorts it into different types of consumer, making its judgement by where the client lives. The sorting first begins by linking the address of each client to a pre-determined classification of the type of area that address is found in. As a consequence, the clients are segmented into groups not actually on the basis of their own, individual characteristics but according to some sort of social average for the area in which they live – by the type of area in which they reside (this distinction is important and one we return to). The area type is defined by the classification used to sort the consumers into groups. Such a classification would normally be purchased from a third-party data vendor (see Chapter 3 for examples). A 'look-up' file then allows the retail company to determine in which type of neighbourhood each of its customers lives.

As we discuss in Chapter 6, the neighbourhood classification is produced by the data vendor as a statistical amalgam of small-area (often census) data for the N mapping units of the region concerned (usually a country or nation). There are, for example, $N = 175,434$ small output areas in England and Wales following the 2001 Census. There are approximately $N = 8.5$ million blocks recorded in the 2000 US Census and there are $N = 38,366$ meshblocks covering New Zealand for its 2001 Census. The geodemographic classification is produced by grouping the N areas into a much smaller number of k classes, on a like-with-like basis. Commonly k is in the range from about 10^1 (10 clusters or area types) to 10^2 (100 clusters or area types), depending on the level of granularity required for the analysis. For example, the 1991 Census-based SuperProfiles classification of the UK has $k = 160$, $k = 40$ and $k = 10$ available to choose from, allowing the user some choice in the precision of geodemographic analysis desired. That hierarchy of clusters is shown in Figure 1.5.

The retail company completes its analysis by comparing the proportion of its clients in each of the k classes with the corresponding proportions for all consumers within the company's catchment area (or some other suitable measure). This comparison of the observed distribution by neighbourhood type with an otherwise expected baseline distribution allows the retail company to infer useful information about its core customers and market to them accordingly (see Chapter 5 and also Birkin, 1995).

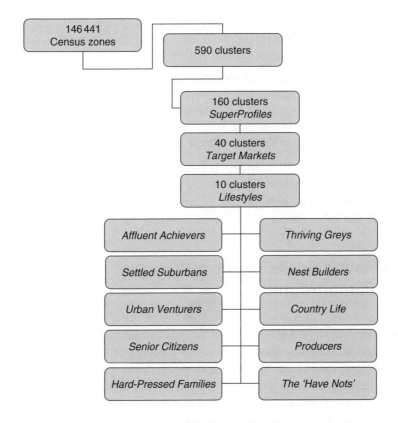

Figure 1.5 The post-1991 Census, SuperProfiles classification hierarchy. Source: based on Batey and Brown (1995, p. 94)

An interesting and somewhat contentious real-world application of the SuperProfiles classification has been by the Higher Education Funding Council for England (HEFCE). It has used geodemographics as part of its programme of widening participation in higher education by increasing the recruitment and retention of students from under-represented social groups. HEFCE (1998, Annex A, paragraph 17; and 1999, paragraphs 24 and 22) make clear that:

> Our preferred method of measuring the socio-economic background of students is through the use of the postcode to identify the neighbourhoods of students' homes. We have found that the Super Profiles geodemographic classifier is able to distinguish between neighbourhood types with markedly different higher education participation rates and, currently, this classifier is our first choice for funding allocations. [para 17]

[T]he additional funding will be applied to young full-time undergraduate entrants to HE [Higher Education] from lower than average participating geodemographic groups in each institution. [para 24]

Although we recognise that there are some flaws in the use of geodemographic data, we believe this is the best measure currently available for allocating funds in respect of students from disadvantaged backgrounds. Therefore, we will use this measure until any better and more precise measures are developed in the future. [para 22]

At the time of writing, an education-specific version of the UK Mosaic geodemographic classification (see Chapter 3) was being developed in conjunction with the UK's Universities and Colleges Admissions Service (UCAS) (Farr and Singleton, 2004).

1.5 How it works: a short theory of geodemographics

In seeking to define the nature of geodemographics, one of us (Harris, 2003, p. 225) has suggested it is 'the analysis of socio-economic and behavioural data about people, to investigate the geographical patterns that structure and are structured by the forms and functions of settlements.' Taking time to consider what that long-winded statement actually means sheds light on why neighbourhood classifications may be useful, predictive tools for analysis and decision taking.

First, the statement suggests that identifying geographical patterns or trends within societies is an important step towards understanding the processes and phenomena that gave rise to those patterns in the first place. Where appropriate, those processes or geographical events can then be managed. Admittedly, it is a simplistic 'theory' and the link between pattern and process is often much more ambiguous. For example: two or more completely different processes can result in similar outcomes (geographical patterns); particular socio-economic processes do not operate in isolation from others present at the same time and in the same place, so disentangling the effects of any one from the others may be difficult, perhaps even pointless; any one outcome may be unique to a particular time and place and, in such a case, ought not be generalized; and, finally, since order can arise from chaos, simple notions of cause and effect may not be tenable (cf. Flake, 1998).

Such caveats are important and increasingly central to geographical and other disciplinary understandings of how settlements and societies form and function (see, for example, Longley and Batty, 2003). However, Webber (1975) suggests that whereas most social scientific research is undertaken in order to investigate existing hypotheses (the deductive approach), one of the key purposes of a geodemographic classification is to generate new ideas and insights that can then become subject to further debate and analysis (an inductive approach; cf. Chapter 9, Section 9.2.2). In this regard, geodemographics is regarded as a data exploration tool, not a statistical method of hypothesis confirmation or rejection. While we fully recognize and will explore a number of limits to the geodemographic method, at the same time we echo the sentiments of an editorial written in response to an article by one of us, in which had been questioned the merits of the geodemographic approach (Harris, 1999). The editor wrote:

> The elegant simplicity of the insights that can still be gained into the overall profile of one's customers by a simple analysis of postcodes is one which continues to impress this practitioner. [. . .] Mapping and geodemographics continue to offer powerful benefits (Whitehead, 1999, pp. 109–10).

If we consider again an analysis of neighbourhood deprivation that is intended to guide the allocation of finances to local regeneration schemes, it is self-evident that targeting resources to the right communities requires that the locations of those communities first should accurately be determined (Policy Action Team 18, 2000). Cross-referencing those locations with, for example, an analysis of house prices might give an indication of the sorts of socio-economic processes (in this case, the effects of housing markets) that cause the geographical differentiation of communities and which lead to social exclusion. The fact that various socio-economic forces can 'filter' people to live in certain 'types' of neighbourhood is the basis for how neighbourhood analysis predicts the likely population characteristics to be found in any given area. Hence, pattern reveals process, and process invites prediction – of, for example, changing consumer markets (Miles, Anderson and Meethan, 2001) or of the growth of urban sprawl (Besussi and Chin, 2003).

However, the relationship between places and people is neither one way nor solely the consequence of external factors. When people speak of 'their neighbourhood' or 'their community' they do so in a way that suggests an attachment to place. Harris (2003) implies that there is an inter-relationship between people and places – the link illustrated by Figure 1.1. Therefore, the physical, social and economic properties of settlements in some way reflects the character, choices, preferences, ideals, affluence, consumer lifestyles (and so forth) of past and present populations living in those settlements but also are a consequence of governmental policies, for example in respect of planning controls and social housing initiatives. Because a place usually pre-dates the residents, so the relationship is two way: the style and character of the settlement 'draws in' certain population groups, perhaps by choice, perhaps by necessity; those residents then shape further evolution of the area. Longley and Batty (1996, p. 76) write that:

> [t]he behaviour of individuals in [geographic] space together contribute to the development of places over time and these place effects in turn condition subsequent spatial [geographical] behaviour.

The interrelationship suggests that measures of the physical, social and economic properties of settlements can yield useful information about the characteristics, preferences and lifestyle choices of the populations resident within those settlements, because people and places are dependent on each other.

These theoretical ideas are summed up by the adage 'birds of a feather flock together'. This, according to Flowerdew and Leventhal (1998) is the basic tenet of geodemographics. In fact, birds of a feather may not just flock together but also increasingly become alike. This is because very few of us (the birds) live in complete isolation from the rest of the society (even if there are times when we wish otherwise!). It is likely that many of our behaviours, choices, aspirations and ideals are influenced by those with whom we interact in our everyday lives (and vice versa) and to assume otherwise is known as the atomistic fallacy (see Chapter 9). Despite the emergence of cyberspace and the popularity of online chat rooms or other forms of communication, it remains reasonable to suppose that geographical distance and location impart constraints on who we meet and when. Weiss (2000, p. 25) argues that there is value in classifying populations at a neighbourhood level and it relates to a 'core truth . . . you are like your neighbors.'

The geographical effects of 'birds' flocking together are expressed by Waldo Tobler's often quoted first law of geography: 'Everything is related to everything else, but near things are more related than those far apart' (Tobler, 1970; see also the forum discussing the law in the *Annals of the Association of American Geographers* **94**(2)). This 'law' is an expression of what spatial statisticians refer to as spatial autocorrelation (Cliff and Ord, 1973). This type of autocorrelation is present in a dataset if it can be shown that particular attributes of the population (such as low-income pensioners, lone-parent households, eat-out regularly couples or sports-car-owning adults) display a non-uniform and non-random patterning but, instead, are clustered into particular localities.

Geodemographic methods assume (positive) spatial autocorrelation when residents of the same neighbourhood are taken to share, in broad terms at least, some common socio-economic and/or behavioural characteristics. This stance permits the consumer profile of a non-sampled person living in a certain neighbourhood to be inferred from data about other persons living in the same neighbourhood – the assumption being that proximity is related to similarity. However, Tobler's first law does need to be modified when looked at in a geodemographic context. The geodemographic methods that are the subject of this book assume not only that proximate populations are related but so also are populations living in the same 'class' of neighbourhood. In other words, near and far things are related – by neighbourhood type.

The use of the word 'neighbourhood' is problematic here because, as we explore in later chapters, it can have formal and informal, administrative and sociological meaning (Martin, 1998b). Moreover, from an 'internal' perspective, Weiss (2000) identifies that while geodemographic literature has often interchanged the words cluster and neighbourhood – for reasons that if not already apparent will become so in Chapter 6 – such synonymy is no longer always appropriate. The reason is that classifications have been built at increasingly fine resolutions. The evolution in the UK, for example, has been first electoral wards, then census small areas, then even smaller unit postcodes and, finally, households and individuals. Intuitively, at some scale between wards and households, the importance of neighbourhood has disappeared – but at which scale?

The answer, we suggest, is none. Unless we believe that people live in a social vacuum, that physical proximity to others has no effect on our own behaviours and that everyone has an entirely unconstrained choice about where they live, then geography remains important and so the presence of 'neighbourhood effects' will be captured even if we collect data at an individual or household scale. Geography matters and is usually there somewhere within the data, not least because apparently individual or household classifications usually incorporate areal data to improve the robustness of those geodemographic models.

These considerations partly explain why we prefer the term neighbourhood over cluster type and use it more frequently (though not exclusively) throughout this book. It is also because the use of the word neighbourhood helps identify the origins of geodemographics in social research (Chapter 2). Finally and most pragmatically, the majority of this book *is* about classifying small areas ('neighbourhoods' in a strictly technical sense) and about using geography as the basis for modelling, even inferring people's demographic, socio-economic and behavioural characteristics. We do not ignore individual or household classifications but the focus is more geographical.

It may also seem ornithological! Returning to birds of a feather, neighbourhood classifications use multivariate (many variable) clustering techniques to model geographies of 'the flocks'. They simplify a complex geographic reality to make the basis and process of decision making easier, faster and more intelligible to stakeholders such as boards of directors or front-line operatives, including store managers and divisional superintendents. Like any other geographic model, geodemographics can both be criticized and praised for the level of abstraction they impart. This is especially true of multivariate techniques that numerically summarize often complex spatial patterning within datasets and consequently create an uneasy, operational tension of trying to express that complexity using a simple and understandable, 'everyday language'. The tension appears most acute when geodemographic vendors attempt to sum up each type of neighbourhood produced by their classification techniques using evocative vocabulary. Unsurprisingly it is these verbal labels or 'pen portraits' that have attracted most criticism within some academic literature (notably Goss, 1995; Curry, 1998). In any case, there are conceptual, theoretical and practical limits to what any area classification or dataset can usefully reveal about the character of populations living within neighbourhoods (we discuss some of these in Chapter 8). Accordingly, there are times when neighbourhood analysis is appropriate and times when it is not. Discerning between these two instances is important and by means of this book we hope to facilitate such a judgement.

One important use of neighbourhood classification is when analysts encounter the task of predicting the likely behaviours and needs of existing or potential new customers or clients in the absence of direct information about those customers, clients or consumers. If the scenario sounds unlikely then consider that for various reasons which include privacy laws and the regulation of personal data, specific personal data about individuals is usually less easily obtainable (and then more expensive to purchase) than aggregate, summary information about groups of people. In some cases, obtaining individual information may actually be illegal! Even supposing that individual-level data were available to the analyst, it is rare for such a source dataset to be fully relevant, comprehensive and complete for the entire population to be analysed (see Chapter 9 for a discussion of these issues in the context of 'lifestyle databases'). Geodemographic information can therefore fill holes in our knowledge base. In the absence of more specific information, knowledge about neighbourhoods provides a useful (and perhaps only) step towards knowledge about people. Combining this knowledge with what is revealed from loyalty card, EPOS (electronic point-of-sale) or other sources of consumer/population data is the first step towards, for example, building predictive models of consumer behaviour or to targeting accurately community funding.

Case study: modelling price sensitivity and geodemographic categories in the restaurant market

Martin Callingham, professor in geography, Birkbeck College, University of London (formerly group manager of market research for Whitbread plc)

This case study is about how geographically to tailor prices within a market. Specifically the work was done in the restaurant market for a chain of about 280 pub restaurants called *Brewers Fayre* that was geographically dispersed about the UK. The sites selected to build these restaurants were semi-rural and the product offering was a low price but substantial, traditional English, full restaurant meal. The chain was very successful (a brand leader) and this appeared to offer potential for increasing the overall prices. However, there was concern that such an increase would negatively impact upon outlets located in poorer areas and this in turn suggested differential pricing should be introduced. To make this introduction, two questions needed to be addressed:

- How to identify the range of prices to be used.
- How to allocate each restaurant to a price band.

To answer these questions a highly novel approach was adopted and used to attribute a price-sensitivity index to ACORN classes of neighbourhood type (ACORN is a commercial geodemographic classification). From existing knowledge of the ACORN mix within each of the restaurants' catchment area it would then be possible to model the collective price sensitivity around each restaurant. However, an important methodological issue had first to be addressed. Specifically, what should be optimized by varying the price? Is it the number of trips to the outlet or the profit that accrues from the trips? Clearly it would be possible to lower prices and increase total trip numbers but this is of no value to the company or its shareholders if the effect is to lower net cash profit. The intent, therefore, was for the restaurants to be more profitable.

The method adopted was to construct a market research survey of 1200 people stratified by neighbourhood type (200 per ACORN category) and where each person was located within the catchment area of an established Brewers Fayre restaurant. To enable the efficient use of interviewers, 30 (about 10%) restaurants were selected at random from the chain list, using an interval sampling method. Interviews, which were conducted face to face, broadly were quota controlled by ACORN category within each catchment area and interviewers were given a postal address list to help them achieve this.

The price-sensitivity questions used were a modification of the method developed by Van Westendorf. This is an unusual method that allows individuals to select prices at their personal conceptual breakpoints. Specifically there are four breakpoints: two 'outside' prices – the price which is so low that the quality of the product would be in doubt and the price at which the product is too expensive – and two 'intermediate' prices at which the product seems either on the cheap side or is becoming too expensive. Respondents were asked to select a price for a typical main meal at a Brewers Faye restaurant from a choice of possible prices covering the entire sensible range (and more!) and broken into a non-obvious interval to avoid conceptual 'rounding up': in this case the prices differed by seven pence units. The meal was described to interviewees using a show card.

The resulting data are normally analysed by a cumulative process that suggests the proportion of the sample that is prepared to pay a particular price for a particular meal. However, in this study a more insightful modelling method was preferred. This was based on asking two additional questions about behaviour at two of the prices given – specifically, the breakpoint prices identified by each respondent (and therefore differing from individual to individual). At the two outside prices, where the product either is too cheap or too expensive, no sale would be expected so the frequency of visiting a restaurant would be zero. However, at the two intermediate prices, sales would be expected, so the interviewee's likely frequency of visiting a restaurant at these two prices was asked. Generally, though not always, the frequency was higher for the lower of the two prices.

The price sensitivity of each individual was then modelled using three linear equations: one for the prices between 'too cheap' and 'on the cheap side'; one for prices between 'on the cheap side' and 'on the expensive side'; and one for prices between 'on the expensive side' and 'too expensive'. Subsequently, the three equations were used to impute an expected visiting frequency of each individual at any specified price. The derived frequency of visiting a restaurant for each respondent was calculated within a fixed range of prices: in this case for 25p differences from £3.50 to £8.50 giving 21 different prices in all (the current price for a main meal was actually £5.40).

Next, the individual frequencies were grouped by ACORN category. This permitted the mean frequency of visiting (and its variance) to be calculated by neighbourhood type, giving a grid of six ACORN categories by 21 imputed frequencies. At each position in the grid the cash profit for a trip to a Brewers Fayre was calculated by applying financial analysis rules to determine profitability. (Profit is generally taken at the outlet level to be the sales less the costs of achieving them but how costs are attributed to sales is quite complicated. Some costs per sale, such as the cost of capital, get less as the sales increase since there are more sales to spread the 'fixed costs' over. Some costs go through steps changes as sales increase – the need to put on more staff, for example – and some costs, such as ingredients, stay very similar per meal sold.)

The resulting (second) grid of net cash profit was then separately multiplied by the ACORN profile of each of the restaurant catchments where the profile expressed the actual number of people of each neighbourhood type that were to be found within each catchment. This gave the modelled net cash profit that could be expected for each restaurant at each of the 21 price points and thereby suggested the most effective profit-generating price for each specific restaurant. A distribution of these 'most effective' prices was made and from this three price bands were selected to be applied across the chain. Each restaurant was allocated to a price band according to the geodemographic profile of its catchment and according to the modelled price sensitivity of that particular socio-economic mix.

Case study: using geodemographics in the public sector

Keith Dugmore, Demographic Decisions Ltd and chair of The Demographics User Group

The development of geodemographics in the public sector of the UK really has been rather odd! Taking the core of the public services to be central government departments, local authorities and health authorities, interest in area classifications based on Census statistics began to develop as early as the 1960s and blossomed in the 1970s. There were several strands. In local government, the Greater London Council (GLC) was very active, producing classifications of boroughs and then electoral wards. Liverpool Council developed a finer scale Enumeration District (ED) level classification focused on deprivation. In central government, the Office of Population Censuses and Surveys published a national classification of local authority districts, while in the mid-1970s work in this field was pushed further forward at the Centre for Environmental Studies (CES), developing nation-wide classifications down to ward level. This early history of geodemographics is discussed in more detail in the following chapter.

Meanwhile, in another part of 'the forest' during the late 1970s, area classifications began to be applied for target marketing. This topic is covered in more detail in Chapter 3 of this book. Suffice to say that starting with ACORN, which provided an ED level classification for the whole of Great Britain, there were two vital innovations. First, the statistical clusters were popularized using names, descriptions and photographs. More importantly, the EDs were matched to the Central Postcode Directory to produce a postcode to geodemographic code look-up table – a means to link a person's property address to their neighbourhood type.

It was this that led to a great leap forward, also enabling area classifications to be linked to other datasets and thus giving birth to geodemographic analysis, rather than just standalone area classifications. Analysis took three main forms and opened up new territory for users in thousands of commercial companies. It also sets the scene for considering geodemographic developments within the public services. These forms are:

- Coding and analysing sample surveys. By taking records of respondents coded with a postal unit identifier and attaching a geodemographic code to these it became possible to profile sample surveys. The most celebrated early example was the contrast between the profiles of *Daily Telegraph* and *Guardian* readers exposed by the Target Group Index survey (a consumer research survey – see Chapter 3).
- Coding and profiling customer files, seeking variations between different segments of the data for such topics as consumption of goods and services, and customer profitability.
- Area analysis, profiling non-standard areas such as postal sectors and store catchments (see Chapter 5). When combined with profiles from surveys or customer files, this opened up the potential for estimating and comparing market potential in different parts of the country.

Such analyses have proven to be of great value to commercial companies. Suppliers have vied to produce new geodemographic classifications and their use has expanded in the last two decades to become second nature in almost all large and medium-sized companies which sell in the consumer marketplace.

Meanwhile, following the arrival of the 1981 and then 1991 Censuses, the public services continued their keen interest in area classifications. But how much geodemographic analysis, linking datasets, was done? The answer is: remarkably little! Taking the three forms of analysis done by commercial companies, there are several examples but these are sporadic.

Some sample surveys have been coded. The Home Office's British Crime Survey was first coded with ACORN back in the 1980s and this led to some fascinating analysis of fear of crime in different types of neighbourhood. More recently, the ONS' Expenditure and Food Survey has been coded with both ACORN and Mosaic but this is another rare exception among the very large number of government surveys.

Turning to the coding of customers or administrative files, progress has been very patchy. Following the report of the Korner Committee on Health Service information (The Korner Committee, 1984), health authorities began during the 1980s systematically to postcode patients' files and this led to some interest in geodemographic coding. Some police forces have also shown interest (although this has often evaporated when the pioneer responsible was promoted). In general, few local authorities have sought to code files and it is significant that the addition of postcode analysis to the ubiquitous SASPAC Census software in 1991 was little used.

As for area analysis, the interest in producing area classifications has continued unabated, with both general and specific (for example health) classifications being produced for a variety of standard administrative areas. In some cases, such as the Index of Multiple Deprivation 2000 and the updated Index of 2004 (DETR, 1998; ODPM, 2004), non-census data sources have also been included but the classifications themselves usually have been treated as free-standing, rather than integrated with survey or administrative data to estimate, for example, the incidence of smoking in small areas.

There appears to be no single reason for this slow progress but the factors probably include the traditional cultural divide and mutual incomprehension between private and public sectors in Britain, manifested in this case by a reluctance to adopt techniques that are used by marketers and also the fact that the commercial geodemographic classifications have not been available free of charge. The result has been that the use of geodemographics has not become everyday in most public-service organizations.

However, this looks set to change. From a policy point of view, governments have become more interested in targeting their resources, both to areas and individuals, and this seems likely to continue. The policy of neighbourhood renewal in the UK has triggered a demand of better data for small areas and the development of the various neighbourhood statistics services (see Chapter 3, Section 3.2). This will increase the use of multiple datasets for small areas and should encourage more creative integration of sources.

Turning to the supply side, the most dramatic change is that the ONS is producing an area classification down to the finest census output area (OA) level for the first time. This is for the whole of the United Kingdom and, when allied with the postcode to OA look-up directory and digital boundaries, which are freely available as part of the Census Access project, will provide an immensely valuable resource. The door is now open to much more adventurous exploration of geodemographic analyses by public services. The most obvious starting point would be for the new classification to be appended to all government surveys and for profiles to be produced as standard products for a wide range of topics such as crime, health and education. Both these and profiles derived from coded administrative files – such as the Inland Revenue – could be included as part of the Neighbourhood Statistics website. The availability of such information from central government would then have every chance of encouraging similar developments among local government, heath authorities and police services.

1.6 Where next? An overview of the following chapters

In this chapter we have introduced geodemographics as the analysis of people by where they live, identified how neighbourhood classifications might be used for both public- and private-sector decision making and offered some ideas on how neighbourhood-based analysis can be a useful way of making sense of geographical information.

In the next chapter we look at the precursors to the present geodemographic industry, in particular Charles Booth's studies of poverty within London and the Chicago School of urban sociologists. Our gaze is more than one of academic curiosity but views also the origins of the geodemographic method, including its strength and weaknesses as an analytical approach. Here we introduce the risk of ecological fallacy, something that has been described as the stick by which geodemographics is beaten. In Chapter 3 we continue our study of the evolution of geodemographics exploring how it came to be commercialized and the 'big players' that have emerged within the sector.

In Chapter 4 we shift our attention sideways to look at another field where analysing geographical information is paramount. Here we outline the principles of geographical information systems (GIS) but do so from the perspective of a geodemographic user. We also make connections between GIS and geodemographics but go on, in Chapter 5, to suggest that GIS are not always necessarily the best software to undertake typical sorts of geodemographic analysis such as neighbourhood profiling or catchment analysis. Therefore, in Chapter 5 we also compare and contrast GIS with what have been labelled geodemographic information systems (GDIS). We describe the sorts of functionality a user might expect from a GDIS and demonstrate the types of analysis that may be carried out within it.

Chapter 6 is about building a neighbourhood classification and considers in detail the data sourcing, validation and grouping strategies employed by commercial vendors, as well as the sorts of issues they encounter when constructing a neighbourhood typology. Chapter 7 considers some of the differences in the construction and geographies of neighbourhood classifications that have been built around the world.

Chapter 8 adopts a more critical perspective and asks – 'but does geodemographics work?' Here some of the shortfalls of the geodemographic method are addressed, notably the problems of diversity within neighbourhood types, modifiable units and whether geographical data require explicitly geographical ways of analysing them. The critique is counterbalanced by the ultimately pragmatic consideration 'well, it has worked for many!' and suggests why. Examples of where geodemographics has proved useful are given.

Chapter 9 brings with it a change of scale. Whereas all the preceding chapters are about top-down, area classifications, here the application of segmentation techniques to individual or household data are considered, together with 'bottom-up' geolifestyle methods. Such approaches require that special consideration be given to personal data protection issues; issues of data uncertainty must also be addressed. Finally in what we have referred to as a postscript, we identify the 'three Is' that gave rise to geodemographics and continue to shape its future.

1.7 Conclusion

Geodemographics is the analysis of people based on a statistical classification of the area in which they live. The classification aims to capture the important socio-economic 'dimensions' of, and differences between, neighbourhoods. The geodemographic approach has been found by many to be a useful aid for guiding decision making and the management of geographical information.

Neighbourhood classifications usually are produced by grouping together a large number of usually administrative units into a much smaller number of groups, clusters or neighbourhood types on a like-with-like basis. A common choice of data to define the similarity, or otherwise, of neighbourhoods are national census statistics. Such classifications have been the bedrock of a rapidly growing industry that has its origins in urban geography and sociology. Present applications include survey design, retail planning, direct marketing and media analysis, as well as other strategic marketing, planning and decision taking in both the public and private sectors. Neighbourhood classifications can be used to look for geographical patterns in various socio-economic, behavioural, attitudinal or consumer datasets.

The usefulness of neighbourhood classifications derives from the idea that knowing where someone lives provides useful information about how someone lives. A simple theory of geodemographics is there is an inter-relationship between people and places, and also between individuals and the people they regularly meet. The adage, 'birds of a feather flock together' and Tobler's 'first law of geography' go some way to explaining why neighbourhood classifications can usefully be applied to extract information about people from information about places. When direct knowledge about potential customers, clients or consumers is not consistent, neighbourhood analysis provides an inferential tool linking what is known to what is not.

Summary

- Geodemographics has been described as the analysis of people by where they live.
- It has been widely used to inform strategic marketing and planning.
- Neighbourhood classifications usually are produced by grouping a large number of administrative units into a smaller number of clusters, on a like-with-like basis.

- An assumption is that 'birds of a feather flock together' such that populations living in the same neighbourhood type share broad socio-economic and consumer characteristics.
- Multivariate classification techniques simplify a complex geographic reality to make the basis and process of decision making easier, faster and more intelligible to stakeholders.
- Geodemographics is better for exploratory analysis than for hypothesis testing.

Further Reading

- Longley, P. and Clarke, G. (eds) (1995) *GIS for Business and Service Planning*, GeoInformation International, Cambridge, Chapters 5–6.
- Sleight, P. (2004) *Targeting Customers: How to Use Geodemographic and Lifestyle Data in Your Business*, World Advertising Research Center, Henley-on-Thames.
- Weiss, M. (2000) *The Clustered World*, Little, Brown, New York.
- Our website: www.geodemographics.info.
- The Geodemographics Knowledge Base: www.geodemographics.org.uk – a comprehensive list of websites for people interested in the application of geodemographics and geospatial analysis, produced by the Census and Geodemographics Group of The Market Research Society.

2
London to Chicago and Back Again! The Origins of Geodemographics

Learning Objectives

In this chapter we will:

- Look at the early history of geodemographics, in particular Charles Booth's maps of London and the Chicago School of urban sociology.

- Discuss how these inform current geodemographic practices, including their strengths and weaknesses.

- Introduce the risk of ecological fallacy and the need to consider population diversity within neighbourhoods.

- Discuss what we mean by neighbourhood.

- Explore the parallels and divergences between commercial geodemographic classifications and governmental measures of deprivation.

- Introduce the third of our expert case studies, entitled 'Charles Booth – yesterday once more?', written by Scott Orford of the Department of City and Regional Planning, University of Cardiff.

Geodemographics, GIS and Neighbourhood Targeting   Richard Harris, Peter Sleight and Richard Webber
© 2005 John Wiley & Sons, Ltd   ISBNs: 0-470-86413-3 (HB); 0-470-86414-1 (PB)

Introduction

'Why,' you might be thinking, 'would anyone other than misguided academics or amateur historians be interested in the evolution of geodemographics?' The answer, of which we hope to convince you, is that knowledge of how present-day neighbourhood classification methods 'came to be', sheds light on the advantages and disadvantages of current practices, and permits a first look at some of the strengths and weaknesses of the geodemographic method, to be explored further throughout this book.

Of course, few if any ideas spring from nowhere; most apparently original thinking builds on preceding ideas or conventions. Consequently, identifying the one true origin of neighbourhood analysis and targeting would be a problematic venture. Fortunately, our objectives here are more limited: our intention is to review early examples of neighbourhood classification and show, by chronological review in this and the following chapter, how these underpin the 'state of play' in today's geodemographics industry. Our focus is less on the historical detail but on the ideas, theories and applications that have interwoven current geodemographic systems and methods of neighbourhood targeting. A particular focus of this chapter is on neighbourhood measures of deprivation and poverty. In the next we focus more of our attention on private-sector applications.

2.1 The life and labours of an early neighbourhood analyst

Published in 1889, Charles Booth's *Descriptive Map of London Poverty* arguably is the first example of applied geodemographics. The history of this map – the 1898–9 revision of which can be viewed online at http://booth.lse.ac.uk/ – begins a few years earlier, specifically the 10 September 1886, when Booth began his research into what would become known as the *Life and Labour of the People of London* (Booth, 1902–3). Seventeen years and 17 volumes later, his enquiries were complete!

What was Booth's motivation for such an extensive and privately funded undertaking? Pfautz (1967, p. 21) argues that the clue is to be found in an autobiography of H. M. Hyndman of the Social Democratic Federation (Hyndman, 1911, p. 303). In the autumn of 1885 the Federation had published the results of a sample study it had conducted, determining that 25% of the workers of London's metropolis earned insufficient wages to prevent 'the slow but sure physical deterioration' of themselves, their wives and their children. Although, Booth had experienced the unhealthy conditions of London neighbourhoods during an unsuccessful contest for a parliamentary seat in 1865, he did not believe that unyielding impoverishment affected as much as one-quarter of the capital's population. According to Pfautz (1967, p. 21),

> [T]his [claim] was too much for Booth's essentially conservative point of view; he felt that the publication of such a figure was incendiary [. . .] Beyond Booth's understandable desire to prove the Socialists wrong in their estimate of the extent of the poverty in London, the urgency of some kind of action was further motivated by the run of events [. . .] [T]he Pall Mall Gazette supported the federation's findings [. . .] 'shortly afterwards, rioting again broke out in the West End, and it was rumoured that the unemployed had been incited to these extremities by Hyndman in person' (citation from Simey and Simey, 1960, p. 69).

Ironically, Booth's own study was to actually indicate 30.7% of London's population as below the poverty line: 6% more than the Federation's original estimate!

Booth began his study at Tower Hamlets in London's East End (Booth, 1887), extending the inquiry a year later to include the people of East London and Hackney (Booth, 1888). Having been encouraged by the support of the press and the public, Booth subsequently turned his attention to gathering data on the rest of the city (Pfautz, 1967, p. 29). It is here that present-day geodemographics emerges! For the London-wide study, an ambitious undertaking was conceived. The outline of each street in London was carefully shaded on a base map to indicate the general socio-economic condition of the residents. The basis of the classification was the reports of the school board visitors to households in each street. These reports contained detailed records, compiled from continuous home visits, of every family with children of school age. The classification scheme is shown in Table 2.1.

Unlike our geodemographic classification of Atlantic City (Chapter 1), where each street was assigned to one and only one neighbourhood type (or class), Booth permitted some streets to be assigned to multiple socio-economic groups, recognizing the population diversity existing in some places:

> [H]ere and there an attempt has been made to give a little more elasticity to the system by combining the colours. Dark blue in especial will frequently be found with a black line upon it, to indicate that great poverty is mixed with something worse; or a red line has been introduced in connection with pink or yellow to show the presence of a middle-class element amongst working class or wealthy surroundings (Booth, 1967, p. 192, see Pfautz, 1967).

Table 2.1 Booth's classification of streets in London by general condition of inhabitants. Source: Booth (1967, pp. 182, 191–2, see Pfautz, 1967) after Booth (1902–3, *Poverty* series)

Colour code	Description (or 'pen portrait')	Class	Description (or 'pen portrait')
Black	The lowest grade (corresponding to Class A), inhabited principally by occasional labourers, loafers and semi-criminals – the elements of disorder	A	The lowest class – occasional labourers, loafers and semi-criminals
Dark blue	Very poor (corresponding to Class B), inhabited principally by casual labourers and others living from hand to mouth	B	The very poor – casual labour, hand-to-mouth existence, chronic want
Light blue	Standard poverty (corresponding to Classes C and D) inhabited principally by those whose earnings are small (say 18 s [shillings – a unit of currency] to 21 s a week for moderate family), whether they are so because of irregularity of work (C) or because of a low rate of pay (D)	C and D	The poor – including alike those whose earnings are small, because of irregularity of employment, and those whose work, though regular, is ill-paid
Purple	Street mixed with poverty (usually C and D with E and F, but including Class B in many cases)	E and F	The regularly employed and fairly paid working class of all grades
Pink	Working class comfort (corresponding to Classes E and F, but containing also a large proportion of the lower middle class of small tradesman and Class G). These people usually keep no servant	G and H	Lower and upper middle class and all above this level
Red	Well-to-do; inhabited by middle-class families who keep one or two servants		
Yellow	Wealthy; hardly found in East London and little found in South London; inhabited by families who keep three or more servants, and whose houses are rated at £100 or more		

In modern terminology, Booth was developing a type of 'fuzzy' classification with a degree of overlap between the classes (see Flowerdew and Leventhal, 1998; Feng and Flowerdew, 1998). Instead of constraining a street to be of a single socio-economic type only, Booth represented some streets to be of a mixed character. In other words, it was possible for a street of households to be part made up of socio-economic type 'A' and partly of socio-economic type 'B'. The 'all-or-nothing' rule of mutual exclusivity that characterizes most modern classifications (where a neighbourhood is designated to be of one type *or* another *but not both*) was not imposed by Booth. By avoiding the rule, Booth hoped to avoid implying a false uniformity within localities (Booth, 1967, p. 193, see Pfautz, 1967):

> [I]n comparing the one set of figures with the other it must be borne in mind that every street is more or less mixed in character, that the Black streets taken together contain some of every class from A to F, or (including the publicans) even G; the same thing is true of the Pink streets also, and even those with a red line are not without a sprinkling of poor people mixed with the better-to-do.

Booth's reminder that socio-economic landscapes rarely consist of perfectly uniform localities is an early warning to the users of neighbourhood classification of assuming otherwise. The warning is pertinent 'in especial' to those interested in area-based deprivation indicators for targeting regeneration funding. Because 'poor' households are not exclusively located in 'poor' neighbourhoods, so area-based policies aimed at improving the life chances of the impoverished or socially excluded risk missing a subset of that population they ideally would seek to reach (Harris and Johnston, 2003). In general terms, the false assumption that knowledge of the general characteristics of a neighbourhood will always yield accurate and precise information about specific individuals or consumers within those neighbourhoods is known as the ecological fallacy. The problem is illustrated by Figure 2.1. In this light, the geodemographic advertising, cited in Chapter 1 and proclaiming 'we know who you are, because we know where you live' might overstate its message and could be entirely wrong.

Yet, for all his cautions and caveats, Booth at heart appears every bit the pragmatist, an advocate of neighbourhood classification, aware that the loss of localized detail is a necessary consequence of establishing benchmarks for generalization:

> At best the graphic expression of an almost infinite complication and endless variety of circumstances, cannot but be very imperfect, and a rainbow of colour could not accomplish it completely. But in order to group and mass our information we need to sink minor differences, and to this end the shades and combination used may be taken as representing so many types of streets inhabited for the most part by the corresponding classes of people.

> [T]here is dark blue and dark blue, light blue streets vary in character, and there are many shades of black; but the likeness far transcends the difference, and those who know well how the poor or vicious live in one district know pretty well how they live in all other districts. I would even venture to say that the conditions of life do not vary much in any of our great cities, and

the picture I shall try to give for the benefit of those who have no such detailed personal experience may, I believe, be taken as applicable in Liverpool or Manchester or Birmingham, and even in the poorer parts of much smaller places.

Booth, then, thought it possible to generalize across urban spaces, sinking 'minor' differences to illuminate better the centrality of social class in the social organization and functioning of a city. Similar sentiments are echoed a century later by Longley and Harris (1999) who contend that 'generalisation is a cornerstone to rational planning policy' and in Chapter 1 we argued that 'geodemographic classifications simplify a complex geographic reality to make the process of decision making easier, faster and often more understandable to stakeholders.' The risk is

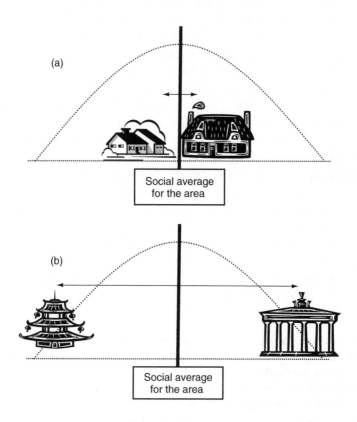

Figure 2.1 We know who you are, because we know where you live? Illustrating the ecological fallacy. In area (a) the specific characteristics of the two households are broadly similar and could be adequately represented by an average index value for the area. In area (b) the characteristics of the two households diverge, so any social average that 'smooths out' the differences will represent neither of them well

that geodemographics oversimplifies, obscuring diversity *within* neighbourhoods while at the same time exaggerating differences *between* neighbourhood groups – an issue we return to in Chapter 8.

Undoubtedly, Booth's poverty studies were the analysis of people by where they live – geodemographics by any other name! An apparent difference between Booth's classification of London streets and latter commercial methods is the source of data: Booth's first classification was based on the opinions of the school board visitors whereas commercial systems frequently are dominated by national census data (Birkin, 1995). Yet, interestingly there has been a trend within the data-marketing industry to make greater use of more unusual and less standard sources of data (in Chapters 6 and 7 of this book we provide examples). Although such information can bring with them issues of uncertainty in the quality and 'representativeness' of the data, it can also be more geographically (and temporally) precise, relevant and interesting than more orthodox sources of information. Such data are often placed under the umbrella term 'lifestyles data' – see, in particular, Chapter 9.

Booth's choice of data necessarily was constrained by what was available at that time. He in fact regarded the 1881 UK Census as entirely unsatisfactory and so became a member of the official committee for the 1891 Census, with recommendations for its improvement! From the results of that later census, Booth developed a less subjective, general classification of the people, albeit one that could only be applied to the less detailed geography of the 127 registration sub-districts in London (see Table 2.2).

Booth also produced an *Index Map of London* to conclude the seventeenth and final volume of his study. Fifty districts were coloured according to an index of their 'comparative social condition':

> The thirty registration districts were the areas [originally] selected, but some of these are inconveniently large and not sufficiently homogenous for useful comparison. By a regrouping of the sub-districts for all of which separate statistics are published by the Register-General, the irregularities have been to some extent obviated, and the whole metropolitan area divided into fifty districts fairly convenient for comparison (Booth, 1902–3, vol. 17, p. 10, cited by Pfautz, 1967, p. 80).

The 50 districts were arranged in five rank orders relative to their average percent of poverty, percent of crowding, birth rate, death rate and rate of early marriage. A global average 'social index' was then calculated based on an averaging of the five ranks. Succinctly, a simple multivariate analysis was applied to census data to produce a classification of

Table 2.2 Booth's "General Classification of the People"

Class		Classification	
Lower Class	1	4 or more persons to each room	Families
	2	3 and under 4 persons to each room	occupying
	3	2 and under 3 persons to each room	less than
	4	1 and under 2 persons to each room	five
Central Class	5	Less than 1 person to each room	rooms
		Families occupying 5 rooms or more	
		without servants	
	a	4 or more persons to 1 servant	Families
Upper Class	b	3 or less with 1 servant	employing
		4 or more with 2 servants	domestic
	c	3 or less with 2 servants	indoor
		5 or more with 3 servants	servants
	d	3 or 4 with 3 servants	" "
		5 or 6 with 4 servants	" "
		7 or more with 5 servants	" "
	e	3 or 4 with 3 servants	" "
		5 or 6 with 5 servants	" "
		7 or more with 6 servants	" "
	f	3 or 4 with 5 servants	" "
		5 or 6 with 6 servants	" "
		6 or 7 with 7 servants	" "
		and other families where number of servants	" "
		about equals that of members of family	
	g	1 or 2 with 5 servants	" "
		3 or 4 with 6 servants	" "
		4 or 5 with 7 servants	" "
		and other families with 8 or more servants, where	" "
		members of family equal the number of servants	
	h	1 or 2 with 6 servants	" "
		1, 2 or 3 persons with 7 servants	" "
		and all families with more than 8 servants, where	" "
		the members of family are less in number	
		than the servants	

Source: Booth (1967, p. 221, see Pfautz, 1967) after Booth (1902–3, *Industry* series).

neighbourhoods by their social condition. Another multivariate methodology is described in Chapter 6 and is the bedrock of present data geodemographics and neighbourhood targeting techniques. And, the more specific method of ranking neighbourhoods according to different factors (or 'domains') of deprivation and then combining those ranks to produce an overall social index is nearly identical to the thinking behind the UK government's *Indices of Deprivation 2000* (DETR, 2000) and *2004* (ODPM, 2004).

A difference between the components of Booth's index and multivariate geodemographic classifications is that Booth's was applied to ordinal not interval data. Ordinal data are ranked and what is lost is the amount of difference between the ranks. For example, it cannot be known that there is a greater change in the poverty conditions from rank 1 to 2 than from rank 2 to 3 unless interval/ratio data are also available which reveal (for instance) that the proportions of poverty are 0.7, 0.5 and 0.45 in the first, second and third ranked areas, respectively. The contrast between interval/ratio measures and ordinal ones are that only the former provide information about *both* the order *and* difference along an assumed scale (cf. O'Sullivan and Unwin, 2003).

2.2 From London to Chicago and beyond!

Pfautz (1967, p. 6) argues that in Booth's classic study '. . . are both theoretical and methodological contributions that make it one of the principal antecedents of the research methods and interests informing the rise of an empirical sociology of the city in America in the twenties . . .' Much of this research took place within the 'Chicago School' (see, in particular, Park, Burgess and McKenzie, 1925) who instituted a paradigmatic and enduring interest among urban sociologists and geographers in establishing general principles about the internal spatial and social structure of cities. These principles were informed by the statistical analysis of social, demographic and economic data, and the focus of the work became on establishing general (statistical) relationships between urban systems and structure, rather than on studying any one particular place or characteristic with more idiosyncratic detail (compare Robson, 1969 or Carter, 1995 with Pile and Thrift, 2000). Significant features of the general urban system or structure could then be summarized using a multidimensional area classification or model, of which perhaps the most famous is Burgess' concentric rings model (Figure 2.2). This modelled the flow of immigration into the centre of Chicago and the movement of existing residents out towards the suburbs, aided by the development of modern, suburban transportation systems to connect them back to their workplaces.

In Figure 2.2, the apparently sharp and fixed boundaries between the various zones of the model are misleading. Burgess actually was describing a process with origins in a biological metaphor of invasion and

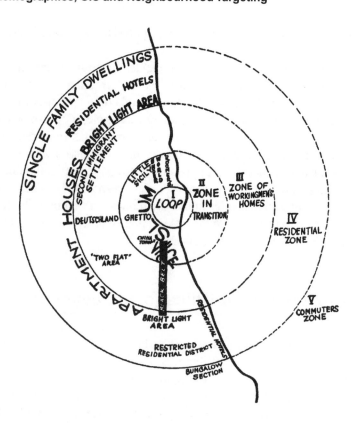

Figure 2.2 Burgess' concentric rings model of the growth of Chicago. Source: Park, Burgess and McKenzie (1925, p. 55), reproduced by permission of The University of Chicago Press

succession. At the risk of mixing metaphors, the rings can be likened to the ripples that flow outwards if you drop a stone into water (the stone is the immigration). The model differs from a geodemographic map of neighbourhood types such as that shown in Plate 1 which is a snapshot in time of certain neighbourhood types in Bristol, England. Plate 1 shows the geographical distribution of the neighbourhood types but is static. It is not intended to convey any particular urban process, although as we discussed in Chapter 1 we might perhaps infer the processes that gave rise to the neighbourhood patterns from the map.

We have recognized that much of the early work in identifying urban spatial structure was carried out for specific American cities, often Chicago. Park's work involved defining 'natural areas' in cities, typifying a field which became known as 'human ecology' (Theodorsen, 1961) – observe again the biological metaphor! Natural areas were conceived as 'geographical units distinguished both by physical individuality and by the social, economic and cultural characteristics of the population' (Gittus, 1964, p. 6). The idea of a natural unit provides one conceptual definition of what we mean when we talk of neighbourhood analysis – the analysis of areal (polygonal) units within which we hope the characteristics of the population are sufficiently similar to each other and also sufficiently different from the populations of other areas that we can reasonably describe each area, and its population, as being and belonging to a neighbourhood; as being a distinct geographical feature.

Set against this object view of neighbourhoods, Kearns and Parkinson (2001, p. 2103) argue that 'there is no single, generalisable interpretation of the neighbourhood.' Although their statement is true in principle (it undoubtedly being the case that different people carry with them different, though not necessarily entirely disparate, mental geographies and therefore different perceptions of what constitutes their neighbourhoods and communities; cf. Jackson, 1989) in policy practice generalized and standardized definitions of neighbourhood are used, albeit that they are usually drawn up for administrative convenience, not to define 'natural areas', express notions of community or to delimit areas of social–economic cohesion. Generalizable interpretations of neighbourhoods exist *de facto*, although there are many of them, specified at a variety of scales: residential addresses; mail delivery units; census enumeration districts; electoral wards; areas of local governance; and so forth. The important point – which we return to in Chapter 8 – is that whichever is chosen for analysing the socio-economic characteristics of population distributions should be fit for purpose. We need to be wary of assuming that the objects of analysis are in any sense 'natural' or necessarily delimit true geographical features in any meaningful way.

Batey and Brown (1995) record that human ecological research was stimulated by the increasing availability of US Census data, issued by census tract (each with a population of about 4000). Local advisory centres helped to define tracts with permanent, recognizable borders, seeking to delimit populations of similar racial and economic status, and residing in similar types of housing. The work of Shevky (Shevky and Williams, 1949; Shevky and Bell, 1955) typified the social area analysis of the time. Three indices were proposed, devised from one to three census variables that measured tract populations on scales of economic, family and ethnic status. The tracts were then sorted into social areas on the basis of their index scores (Berry and Horton, 1970, p. 314). Extension of the analysis to a wider range of cities revealed shortcomings in the original choice of census variables. Subsequently, a greater number of variables were included and the multivariate technique, factor analysis applied – spawning the phrase 'factorial ecology'.

In Britain, early studies of urban spatial structure were hampered by a lack of census data at fine geographical scales (Batey and Brown, 1995). Small-area data became available following the 1951 Census at the level of the enumeration district (ED) – the administrative units that together form the geography of data collection for UK censuses. By 1969 the Centre for Urban Studies (University College London) had experimented with 1961 Census data at the political ward and more detailed, ED level (Norman, 1969). A six-fold classification of EDs was produced for inner London using various data-reduction techniques (specifically, principal components analysis and a least-square cluster analysis method) (Howard, 1969). Batey and Brown (1995, p. 82) show an example of this classification for the Camden area of London, noting the naming of the clusters (Upper Class, Bed Sitter, Poor, Stable Working Class, Local Authority Housing, Almost Suburban). Each class of ED can be compared to the others and to inner London as a whole, against 28 different census variables, of which 14 are shown in Table 2.3. For example, proportionally a greater than inner London average of children aged under five were to be found living in the Poor and Working Class EDs. Such a finding has resonance in ambitious present-day desires to eradicate childhood poverty in the UK.

Foreshadowing the future use of geodemographics in policy making and management, the 1960s witnessed an increasing interest in small-area classifications from local authorities. For example, the Liverpool Social Malaise Study (Liverpool City Council, 1969) sought to identify areas in the city with concentrations of social problems. It is also an early example of the use of data fusion and 'joined-up government' whereby the classification based on the 1961 Census data was then linked to social statistics routinely collected by each council department. (The non census data sources were not used to build the classification, only to interpret it; Webber, 1975.)

The Chicago, inner London and Liverpool studies are pioneering examples of small-area analysis. However, they were also restricted to relatively limited geographical regions. During the 1970s, one of us, working at the Centre for Environmental Studies, London, began to develop national classifications at the ward, parish and local authority levels that permitted clusters of EDs to be compared with the *national* mean for particular census variables (Webber, 1977; Webber, 1978; Webber and Craig, 1978). The classification, subsequently acquired by the company CACI and rebranded ACORN, was then linked to the mail delivery (postcode) geography, enabling it to be used in the private sector to discriminate between the different types of consumer behaviours associated with different neighbourhood classes. Batey and Brown (1995, p. 85) record that 'significantly

Table 2.3 Selected mean characteristics of six types of enumeration districts in inner London, 1961 Census. Source: adapted from Norman (1969, p. 393)

	Classification of enumeration district						All inner London	
	Upper class	Bed-sitter	Poor	Working class	Local authority housing	Almost suburban	Mean	Standard deviation
% population under 5	4.0	4.2	9.8	7.8	6.9	6.0	7.0	2.6
% population under 15	11.2	9.9	21.1	20.1	25.6	17.7	19.2	6.0
% population 65 and over	14.2	11.1	9.8	12.2	9.1		11.8	4.0
Female : male ratio, all ages	1462	1175	1030	1082	1059	1149	1123	237
Female : male ratio, 25–44	1265	919	874	951	1066	1055	1005	226
% women aged 20–24 never married	22.7	26.1	60.6	63.6	43.4	53.1	50.9	20.2
% adults single	40.0	47.8	29.0	25.2	25.4	24.9	28.7	8.6
% one-person households	33.5	46.1	25.2	21.1	13.0	17.4	22.5	11.5
% one- and two-person households	66.0	75.7	53.3	52.2	37.3	50.8	52.6	13.2
% households of five or more persons	8.4	5.4	12.0	11.7	19.7	11.4	12.4	6.2
% all households overcrowded	3.4	11.0	15.7	6.3	5.9	2.2	6.9	5.7
...	:	:	:	:	:	:	:	:
% early school leavers	33.9	40.5	85.1	84.0	86.6	69.7	73.5	21.3
% males in professional/managerial occupations	43.4	25.2	6.6	7.2	5.2	16.9	13.2	14.6
% males in manual occupation	28.7	41.1	77.6	75.2	77.1	55.3	65.3	21.2

better discrimination was achieved, for example in newspaper readership, than had been obtained previously using conventional indicators such as age or social class.' The results of this experimental work (Baker, Bermingham and McDonald, 1979) were presented at the Market Research Society Conference. The presentation was 'the [UK] marketing community's first exposure to this promising new geodemographic approach.'

The seeds had been sown for a new industry to grow and we discuss the commercial development of geodemographics further in the following chapter. It was now possible to compare one census area against another in terms of their neighbourhood profile. However, this offered only limited analytical flexibility, being bound by the geography of the census. What if the user wanted to reach the next stage and undertake the sort of procedure described in Chapter 1: to take a specific client list and sort it into different types of consumer based on the class of neighbourhood within which each client lived?

One further catalyst was required for the geographic industry to blossom: the ability to link people to places (see Figure 1.1, Chapter 1). This has been achieved with the growth of the geographic information industry and increasingly precise and accurate methods to 'tie' specific geographic features to particular geographical locations – methods that include GPS (the global positioning system). In fact, the growth of geodemographics has paralleled the rapid rise of geographic information-handling technologies, notably GIS, which primarily we discuss in Chapter 4.

2.3 A note on measuring deprivation

In this chapter we have observed that much of the early development of neighbourhood classification has been associated with a desire to measure, map and understand geographical patterns of urban deprivation and poverty. This is by no means to imply that deprivation and issues of social exclusion are solely an urban concern, only to recognize that the pioneering studies which nurtured geodemographics were concerned with understanding the built and social structures of cities.

Deprivation is an elusive concept to define and measure. Different conceptions lead to different geographical representations of physical and social conditions, and any classification of deprived neighbourhoods inevitably will reflect the priorities, ideologies, beliefs, knowledge and so forth of those who undertake the classification; the broader social, economic and political climate of the time; and also the technological and data infrastructure available, within which to undertake the classification. There is neither a 'pure object' that is indisputably deprivation nor any perfectly objective and apolitical measures independent of prevailing interests and value systems. (The UN's definition of absolute poverty is a possible exception, giving lack of access to food, clean water, health, shelter and education as the indicators.)

We are not suggesting that deprivation or related notions of poverty and social exclusion are socially constructed to the degree that they have no physical (real) existence – to do so would be an academic indulgence extraordinarily dismissive of the circumstances of those who actually are deprived – only that the methods and understandings used to conceptualize, measure and to target the 'deprived' or, indeed, any other expression of neighbourhood condition are imbued with all sorts of subjective (and hence contestable) social meaning, theory and praxis. The way deprivation is conceived conditions the way that it is measured and, according to the Greater London Authority (Mayor of London, 2002, p. xvi) 'there is no consensus about what is meant by deprivation or poverty, or about the best methods or indicators used to measure them' (although see Lee, Murie and Gordon, 1995). It follows that there is likely to be considerable debate concerning what might be the policy interventions that are most likely to be effective in different, highly deprived localities.

Uncertainties about how best to profile neighbourhood conditions are not necessarily a weakness. To the contrary, they can enable different researchers and practitioners to tailor a solution to the problem. However, they can also give rise to dispute! Consider, for example, Table 2.4, which shows the proportion of the most deprived electoral wards in

Table 2.4 Proportion of the most deprived wards in England that are also within London according to three governmental measures

	Percentage of London wards falling into the 10% most deprived in England	Percentage of London wards falling into the 5% most deprived in England
Index of Local Conditions (DoE, 1994)	46	35
Index of Local Deprivation (DETR, 1998)	40	26
Index of Multiple Deprivation (DETR, 2000)	20	10

Source: Mayor of London, 2002, p. 102.

England that are also within London, defined by various governmental measures. It could be concluded from the table that between 1998 and 2001 the proportion of the most nationally deprived wards at least halved within London. Possible but unlikely! What is more likely is that the different indices vary too greatly in their construction and sources of data for one index sensibly to be compared against another. The difficulty for London's authorities is that the newest of the indices – the Index of Multiple Deprivation (IMD) 2000 – is, at the time of writing, the one used to allocated neighbourhood regeneration monies, with the principal criterion for receiving funding being whether a ward is one of the 10% most deprived. (The IMD 2000 has just been superseded by the IMD 2004; ODPM, 2004.)

Despite the fact that different indices use different variables in different ways, there has still been some agreement about the most important correlates of deprivation (and/or poverty). Common measures used in the UK include levels of unemployment, households that live in overcrowded properties, single-parent households and households not owning a car (Lee, Murie and Gordon, 1995). As we explained in Chapter 1, these and other indicators of deprivation are usually combined in such a way as to give a multivariate profile or score indicating the level of deprivation in a given neighbourhood. If we denote $x_1, x_2, x_3, \ldots, x_n$ as the predictor or independent variables to be included in an index (for example, x_1 could be the proportion of households not owning a car per neighbourhood), then the overall score (y') can be denoted as a function of the predictor variables:

$$y'f(x_1, x_2, x_3, \ldots, x_n) \tag{2.1}$$

For example, the Breadline Britain index (Gordon and Forrest, 1995; Lee, Murie and Gordon, 1995; Gordon and Pantazis, 1997a, 1997b) estimates the percentage of poor households for 1991 Census wards and districts as:

$$y' = 0.2025x_1 + 0.2174x_2 + 0.1597x_3 + 0.1585x_4 \\ + 0.0943x_5 + 0.1079x_6 \tag{2.2}$$

where y' is the estimated percentage of poor households (per census ward or district) and x_1 to x_6 refer to the following 1991 Census variables, respectively: % households not owner-occupied; % households with no car; % lone-parent households; % households where household head is of partly skilled or unskilled occupation; % economically active population unemployed; and % households containing a person with a limiting long-term illness. Note (in Equation (2.2)) that the designers of the index gave increased weight (importance) to, for example, 'households with no car' relative to 'economically active population unemployed' as an indicator of poverty. In Equation (2.2) we have followed convention and used y and x_1 to x_6 to denote the dependent and independent variables, respectively. Unfortunately this also risks confusion when x and y are used to denote geographical coordinates in a GIS (see Chapter 4).

There are many methodological similarities between neighbourhood classifications used for measuring deprivation and commercial classifications used for consumer profiling. Both attach multivariate profiles to neighbourhoods. Both often draw on governmental, particularly census data to help define the profiles. And, for both, what the profiles come to represent exists at the interplay between the real world and the set of methods and meanings used to make sense of it.

However, there are also differences. Most obviously, consumer profiling is interested in accurately identifying the characteristics of different sorts of consumer, while deprivation classifications seek to identify deprived neighbourhoods and their residents. Less obviously, commercial classifications are usually formed by grouping neighbourhoods into clusters or types on a like-with-like basis, with the likeness being defined by the multivariate profile (Chapter 6). Deprivation profiles tend to be ranked from most to least poor neighbourhood and are then essentially sorted into two types, above or below what is ultimately an arbitrary threshold (for example, the top 10% most deprived) – neighbourhoods that are treated as deprived and neighbourhoods that are not. A third difference is in the end use. Many interventions that governments make, such as raising standards in schools or recruiting more police personnel, are targeted on broad geographical areas such as electoral wards, local authority districts, police constabularies or journey-to-work zones. By contrast the programs which direct marketers employ (but less so retailers) usually take place at the individual or household level and therefore make use of data for finer geographical scales.

Case study: Charles Booth – yesterday once more?

Scott Orford, Lecturer in GIS and Spatial Analysis, Department of City and Regional Planning, University of Cardiff

Charles Booth is probably best remembered for his colourful and innovative maps that accompanied his groundbreaking research on the 'Life and Labour of the People of London'. These maps vividly illustrated the social characteristics of the inhabitants of late Victorian inner London for a variety of circumstances and spatial scales. Perhaps his most famous map – and the one that interests us here – is his *Descriptive Map of London Poverty*. This large-scale map shows the streets of inner London, building by building, each one coloured to correspond to one of seven categories reflecting the condition of poverty of the resident household. The seven categories used on the map are described in Table 2.1 (above).

The classification scheme designed by Booth resulted in a surprising amount of discrimination between different household types across very small geographical areas. For instance, the neighbouring areas of Westminster and Pimlico have streets classified using every single category in his scheme: from the 'elements of disorder' around Great Peter Street near Westminster Abbey, through the mixed streets of poverty and affluence in Pimlico, to the wealthy yellow coloured main roads of Belgrave Road, St George's Road and Victoria Street.

This situation is not so different from that of today, with numerous small areas in inner London continuing to demonstrate a wide diversity of social characteristics, and with the wealthy and the poor still living in close proximity. In order to investigate the extent to which the social landscape has changed since Booth's day, Booth's poverty map was digitized and georeferenced to the Ordnance Survey UK's (OS) National Grid coordinate system. Because the Registrar General's social class scheme used in the 1991 UK Census of population is in part derived from Booth's work (and based on five categories of occupation), it is possible to compare Booth's data with 1991 census data.

A GIS (see Chapter 4) was used to construct poverty indices for the digitized Booth data and occupation data from the 1991 Census statistics, with 1991 Census wards used as the comparative spatial unit. The poverty indices are akin to dissimilarity indices used in the analysis of social segregation in the sense that they measure the incidence of relative poverty within each ward by taking into account the relative mix of social classes. Although poverty in its absolute sense has changed beyond all recognition since Booth's day, the degree of *relative* poverty and affluence within each area may be much more stable; inequalities may remain.

Comparing the ward poverty indices reveals some very interesting insights into the changing nature of the social hierarchy of small areas of London. There is a very strong correlation between the two indexes (a Pearson correlation, r of $+0.73$, with no more than 1%

likelihood, p of the association being due to chance), implying a close 'fit' between areas of relative poverty in Booth's time and the same areas in 1991. However, the indices also reveal that, on average, there is less social polarization (between the very rich and the very poor) in 1991 than in Booth's time. Areas in inner London have generally converged over the last 100 years, with less number of areas classified at the extremes of rich or poor and with a corresponding growth in middle-income areas.

A more detailed description of changes in ward-level poverty may be gained by examining how wards have changed in their relative position, ranked from rich to poor. Table 2.5 compares quartiles of wards based on the two ward poverty indexes, one being the richest quartile (the 'top' 25%) and four being the poorest (the 'bottom' 25%). It can be seen that almost half of the wards (46%) have not changed their relative position with respect to poverty between Booth's time and 1991 (the middle diagonal). The greatest shifts in relative position occur in the middle quartiles with very little change at the two extremes, particularly with wards that were richest in Booth's time. The table shows that three-quarters of the most affluent quartile wards have maintained their premier position in the social hierarchy throughout the 100-year period. This compares to the wards in the poorest quartile of which just under half (45%) moved up to a richer quartile by 1991.

A more dramatic illustration of how Booth's classification of inner London continues to have resonance today is provided by examining the correlations between the Booth poverty index and a variety of contemporary standardized mortality ratios for different causes of death. Table 2.6 shows the simple and partial correlations between the poverty indices and the standardized mortality ratios. The Booth poverty index contributed more to predicting deaths from stroke and stomach cancer in

Table 2.5 A comparison of the Booth and the 1991 Census ward poverty indexes by quartiles

		Number and percentage of wards in each quartile			
Quartile		**1991 census data**			
		1	**2**	**3**	**4**
Booth	1	25 (76%)	5 (15%)	2 (6%)	1 (3%)
map	2	6 (18%)	11 (33%)	14 (42%)	2 (6%)
data	3	2 (6%)	12 (36%)	7 (21%)	12 (36%)
	4	0 (0%)	5 (15%)	10 (30%)	18 (55%)

Table 2.6 Correlations between poverty in 1896 and 1991 and standardized mortality ratios for all ages for death in 1991–95

	Simple correlation Booth	Partial correlation Booth	Simple correlation 1991	Partial correlation 1991
All causes	0.56	0.22	0.60	0.35
Coronary heart disease	0.58	0.21	0.65	0.41
Stroke	0.40	0.22	0.36	0.11*
All cardiovascular disease	0.56	0.20	0.61	0.37
Chronic obstructive pulmonary disease	0.58	0.24	0.61	0.35
Pneumonia	0.26	0.07*	0.30	0.17
Lung cancer	0.61	0.30	0.62	0.33
Stomach cancer	0.49	0.24	0.47	0.20

*Insignificant at 5% level.

the late twentieth century than that derived from the 1991 census! For other causes of death, the modern index contributed more. The results of further correlation analyses suggest that for deaths under the age of 65 the 1991 index makes a slightly greater contribution to predicting all cause mortality ($r = 0.56$; $p = 0.01$) than does the Booth index ($r = 0.46$; $p = 0.01$). When only deaths at ages greater than 65 are considered, however, both indexes make a similar contribution to the model: the correlation coefficients are $r = 0.56$ and $r = 0.57$, respectively ($p = 0.01$ in both cases).

The fact that the Booth poverty index performs so strongly as a predictor of mortality is perhaps partly because the median age of death of the people dying in the period 1991–5 is approximately 78. This means that, while very few would have been alive at the time Booth drew his map, almost half would have been born before 1915. The majority of these individuals, however, will have migrated in the intervening period. Hence the predictive power of the Booth poverty index demonstrates how the nature and social hierarchy of different areas of London has remained stable, despite constant changeover of the resident individuals. Surprisingly then, Charles Booth's classification of Victorian London still produces a significant amount of discrimination of the social composition of inner London today. It would appear that wholesale shifts of neighbourhoods within the social hierarchy of the city are a rare occurrence, even over a 100-year time period. It is perhaps this stability that leads geodemographics 'to work' (see also Orford et al., 2002).

2.4 Conclusion

In this chapter we have charted the origins of geodemographics in Charles Booth's descriptive maps of poverty in nineteenth-/twentieth-century London and in the Chicago School's studies of 'the city'. The history we have given necessarily is selective and incomplete but has provided opportunity to discuss some methods of geodemographic study – notably multivariate analysis of national census data and its use in the measurement of social conditions within 'neighbourhoods'. Whether neighbourhood is 'packaged' more in terms of natural units or as formal administrative zones depends on the choice of dataset and the scale at which it is specified. Whether the distinction matters, largely depends on the application – it is, for instance, of more importance when considering where to target community funding than it is for using geodemographic classifications to guide direct marketing communications.

In the next chapter we explore the private-sector evolution of geodemographics and show how its growth parallels the increased, more widespread and ease of availability of geographical information, and of low-cost technological solutions to store, manipulate and analyse such data. We examine the close relationship between geodemographics and GIS further in Chapter 4, but then go on to look at the different software needs of GIS and geodemographic users in Chapter 5.

Before doing so, we may add a 'footnote' regarding the position of geodemographic methods within UK (urban) geography. As we have seen, a range of multivariate, social area and factor analyses were playing a pivotal role in enhancing understandings of the spatial and social structure of cities during the 1960s and 70s. However, by the 1980s interest in this approach had largely waned within academic quarters. The reasons for this are complex but include: a rising interest in the contingent and the particular, and how these 'shape' urban settlements and lifestyles; unease with applying quasi-scientific methods of research to understanding social systems and human agency; the apparent decoupling of analysis from theory, with concern that the methods were primarily data led and technique driven; an emerging political climate that tended to eschew analytical and evidence based policy making; and increased awareness that the ways geographical data are aggregated into reporting areas have effects on analysis that cannot be ignored – most famously, Openshaw and Taylor (1979) showed that the same data could be aggregated in different ways, producing correlations between variables spanning the entire range from -1.0 to $+1.0$!

Ironically, the movement away from geodemographics within urban geography coincided with the acceptance, then development of the methods within a commercial arena. Fortunately, the disengagement has never been absolute. A number of geographers have applied their skills successfully, particularly to the retail sector and, in recent time, there has been something of a rapprochement. This, coupled with the proliferation of geographical data and their associated technologies, has usefully led to some 'old' ideas being revisited and new ones developed. An example of the latter is a recent classification of British surnames which helps to identify how names originate in different locations and reveals some of the historic patterns in the movement of national and cultural groups (Lloyd, Webber and Longley, 2004; see www.casa.ucl.ac.uk/surnames/). There is also increasing interest in using advanced geosimulation techniques to model and study complex urban systems (Benenson and Torrens, 2004).

Summary

- The historical precursors to present geodemographics include Booth's studies of deprivation within London and the Chicago School of urban sociology.
- The development of human ecology, social area analysis and factor ecology was aided by the increasing availability of census datasets and the development of multivariate methods to analyse them.
- In the same way, modern data-handling technologies and the proliferation of geographical data are offering new opportunities to further develop geodemographic methodologies and insight.
- Booth initially developed a type of 'fuzzy' classification and warns that socio-economic landscapes rarely consist of entirely uniform localities.
- The ecological fallacy can be understood as making an (inappropriate) assumption that any specific individual necessarily shares the general characteristics of her or his neighbourhood and its population.
- Formal, administrative definitions of neighbourhood delimit 'natural' and homogeneous groupings of the population to a greater and lesser extent, dependent on scale, location and the purpose of their design. The more 'artificial' the neighbourhood unit the greater the risk of ecological fallacy.

- The risk of committing the ecological fallacy is greatest in areas of population diversity and so varies geographically. The severity of the risk needs to be judged with respect to the context and aims of the analysis.

Further Reading

- Batey, P. and Brown, P. (1995) In *GIS for Business and Service Planning* (eds, Longley, P. and Clarke, G.) GeoInformation International, Cambridge, pp. 77–103.
- The online Charles Booth Archive, http://booth.lse.ac.uk, hosted by the London School of Economics and Political Science.
- The Center for Spatially Integrated Social Science at the University of California, Santa Barbara provides overviews of some of the classic papers and research in spatial analysis – including those of Charles Booth and of the Chicago School: http://www.csiss.org/classics/.

3
The Evolution of Geodemographics and the Market Today

Learning Objectives

In this chapter we will:

- Explore the evolution and globalization of the geodemographic industry.

- Look at how that growth has been characterized by periods of innovation and merger.

- Show how the development of new and updated products traditionally has been linked to the periodic release of small-area census data.

- Identify some alternative sources of data that increasingly are incorporated within or linked to geodemographic classifications.

- Use the UK as a case study to identify more generic developments within the geodemographic industry.

- Present an overview of the US market for geodemographics, written by Dave Miller, senior vice president, Data Research and Development, Claritas USA.

Geodemographics, GIS and Neighbourhood Targeting   Richard Harris, Peter Sleight and Richard Webber
© 2005 John Wiley & Sons, Ltd   ISBNs: 0-470-86413-3 (HB); 0-470-86414-1 (PB)

Introduction

In the preceding chapter we looked at some of the historical antecedents to the present-day geodemographic industry. The foundations of the industry were found to be in academic and social research, so how did it cross that sometimes 'great divide' between business and academia?

In this chapter we explore how a set of specialist suppliers have come to develop the set of geodemographic tools and services that are now used by leading consumer brands and organizations around the world. Our experiences and working lives predominantly have been associated with the UK market and, accordingly, it is there where we focus attention. However, there are a number of more generic observations that will be of interest to an international audience. These are:

- Mergers within and rationalization of the industry.
- The emergence of the 'big players' operating within an international market.
- Yet also new opportunities for entrepreneurialism and smaller 'start-up' firms (often led by former staff of the bigger companies), facilitated by reduced cost access to neighbourhood and other data.
- The broadening of classifications to include non-census based sources of information.
- The emergence of data warehouses storing information not on neighbourhoods but on specific individuals and/or their households.
- The parallel development of geographical information technologies as a catalyst to the more widespread collection, dissemination and analysis of various geodemographic data, as well as to the continued evolution of the industry and the development of new products, supported by new national spatial data infrastructures.
- An increased concern with data privacy and protection (which we consider in more detail in Chapter 9).
- The sorts of business models and relationships offered by vendors to the users of geodemographic systems.

We also include a review of the 'state of play' in the US market for geodemographics.

3.1 From census to commerce

In Chapter 2 we left the story of geodemographics at the Centre for Environmental Studies and its studies of deprivation within Merseyside and North London (Webber, 1977; Webber and Craig, 1978). At the Centre, cluster analysis of census variables had been used to show that the type of deprivation in different neighbourhoods was itself different in nature, requiring different remedies. This work probably would never have come to the attention of the commercial sector had it not been for a combination of circumstances.

While one of us was addressing a conference about our experience of census analysis, an executive from the British Market Research Bureau (BMRB) happened to be in the audience and was interested in an electoral ward-level classification as a possible tool for improving the sample frame used by BMRB's Target Group Index (TGI) survey – a survey investigating consumption of various products, services, brands and media. Subsequently, BMRB coded the TGI survey with the ward classification codes and found interesting differences between neighbourhood types. A paper was presented to the Market Research Society Annual Conference in March 1979 (Baker, Bermingham and McDonald, 1979), which excited considerable interest in the marketing, media and research communities.

It was media research where the greatest initial impact was felt. The often-quoted finding was that the neighbourhood classification was better able to differentiate between *Guardian* and *Telegraph* readers than 'standard' demographics. In terms of age, sex and social grade, there was little difference between readers of the two publications (except that *Guardian* readers were somewhat younger, on average). However, when analysing by neighbourhood, whereas *Guardian* readers tended to be 'inner-city high status', *Telegraph* readers were predominantly 'suburban high status'.

Apart from BMRB noticing the potential of geodemographics, another organization developing an interest was CACI, a US-based company. CACI had set up an office in London in the mid-1970s to offer a service, SITE, whereby retailers could obtain online access to census statistics summarized for bespoke catchment areas. The potential of the electoral ward-level classification was spotted and rebranded as ACORN (A Classification of Residential Neighbourhoods). The UK commercial geodemographics industry was off and running! A similar development was occurring in the USA with the launch of PRIZM (Potential Rating Index for Zip Markets) by Claritas in 1978 under the direction of

Jonathan Robbin, the company's founder (Claritas had also introduced a census-based cluster system in 1947: see Monmonier, 2002, p. 146).

CACI soon produced an ACORN classification at the more detailed, census output area level and linked to the SITE system. (At the time the UK census output geography was the same as the input geography of data collection and output areas, OAs, were actually referred to as enumeration districts, EDs. The separation out of these two geographies occurred post-2001 with recognition that the two should be designed for different purposes – one for efficient census administration, the other for meaningful census analysis. See Martin, 1998a, 2000.) Two milestone developments followed. The first was the ability to link the somewhat opaque census geography to the more widely used and understood geography of mail delivery – unit postcodes (Raper, Rhind and Shepherd, 1992) (these being broadly similar in purpose to the ZIP codes used in the USA).

In GIS parlance, the process of linking the census and postal geographies involved a proximity analysis that related a point grid reference near the centre of any given postcode (the centroid) to the nearest point grid reference representing the centre of an enumeration district. The postcode centroids were provided by a file variously known as POSTZON and the Central Postcode Directory (CPD), originally developed for a regional highway traffic model. The accuracy of the matching process was limited by the technology and geographical datasets of the period. In particular, the resolution (the greatest possible level of detail) of the postcode centroid coordinates was only 100 m, with the assumed centre of each postcode being offset to the south-west corner of the 100 m × 100 m National Grid cell in which it fell (Martin, 1992). Nevertheless, if postcodes could be linked to census output areas then so too could a person's property address (via the postcode), making profiling of customers and cross-tabulation of market research respondents much easier.

With each successive census that linkage has become stronger! From the 1981 Census onwards in Scotland the postcode boundaries had been digitized and the postcodes of census respondents had been recorded, permitting a more accurate, population-weighted match of postcode to census geography. For the 2001 Census of England and Wales the output geography was thoroughly redesigned, with the OAs being specifically built as aggregations (groups) of contiguous unit postcodes and offering, for the first time, near congruence of postal and census geographies at the OA scale (previously, postcodes could be split across census boundaries adding imprecision to the matching process). Over the same time the resolution credited to unit postcode centroids has increased one hundred fold, reflecting more general trends in the evolution of geographical information technologies and data handling. At the time of writing, Ordnance Survey UK was offering their Code-Point product with a 1 m coordinate resolution for postcode centroids.

The second milestone development was an agreement between CACI and CCN (now Experian) to link the former's ACORN classification to the latter's computerized electoral register. The electoral register of Great Britain offers (in principle) the names and addresses of all adults resident in England, Wales and Scotland (there is also a Register for Northern Ireland) – since it is a legal requirement to register to vote – and, at the time, there was no restriction on who could buy a digital copy of the full list of voters. The agreement enabled the Register to be queried by ACORN type, leading to many mass mailshots. In retrospect, this was better news for the geodemographic and direct mail industries than it was for the many recipients! Eventually agencies such as the Mailing Preference Service (www.mpsonline.org.uk) emerged to help prevent unsolicited mail and the geodemographic market developed with two main application areas: direct marketing; and retail, or quasi-retail. Quasi-retail includes applications such as the door-to-door distribution of leaflets and samples, which will often take (for example) the characteristics of the grocery, retail catchment areas into account.

For the four years from 1979 to 1983, CACI *was* the UK geodemographics market! However, competition emerged during the remainder of that decade. Pinpoint Analysis opened for business in January 1983, concentrating initially on applications of census data that did not involve multivariate classifications: for example, the KIDS targeting system (targeting households with high concentrations of children, using 'raw' census variables). But when no competitor to ACORN had emerged by mid-1984, Pinpoint decided to produce its own, launching PiN (Pinpoint identified Neighbourhoods).

The mid-1980s saw the geodemographic market liven up considerably. Webber left CACI in 1985 and joined CCN, which then launched Mosaic in early 1986. Mosaic introduced non-census data, such as credit activity counts and county court judgements into the otherwise census-based classification. At much the same time, SuperProfiles was launched. SuperProfiles' initial development was conducted at the University of Newcastle by Stan Openshaw and colleagues. Details of the methods used to create the classification were published in the academic literature (Charlton, Openshaw and Wymer, 1985) and subtitled 'a poor man's ACORN'! The paper provides an excellent (and more technical) overview of the sorts of classification methods that we also consider, in Chapter 6. Members of the University of Liverpool also contributed to the commercialization of

SuperProfiles. Much of that history is recorded by Batey and Brown (1995). Openshaw went on to propose a number of alternatives to the 'conventional' methods of geodemographic classification (Openshaw, 1989; Blake and Openshaw, 1995; Openshaw and Wymer, 1995; Openshaw, Blake and Wymer, 1995; Openshaw and Turton, 1996).

SuperProfiles was launched onto the market by a subsidiary of OE McIntyre but was swiftly acquired and marketed by CDMS, a Littlewoods subsidiary. Meanwhile, Pinpoint had developed and launched FiNPiN (Financial PiN), using National Opinion Poll's (NOP's) Financial Research Survey data in its construction – the first 'market-specific' classification. Indeed, in the mid-1980s the financial services sector was by far the fastest growing sector of the UK geodemographics market, spurred on by the Financial Services Act that had opened up the financial market to greater competition. By early 1987, there were five neighbourhood classifications jostling for market share: ACORN; PiN; FiNPiN; Mosaic and SuperProfiles.

The emergence of desktop computing technologies paved the way for the development of PC-based market analysis systems in the late 1980s. Prior to this, standard reports and maps had either been obtained using a computer bureau or via a telephone ordering service. Alternatively, postcode directories of classifications could be purchased to enable clients to append geodemographic codes to their own customer records. First from the blocks was Sales Performance Analysis (SPA), which developed a product branded 'The Marketing Machine'. In autumn 1987, CCN bought out the rights to The Marketing Machine from SPA and, linking it to Mosaic, rebranded it 'Mosaic Systems' (subsequently Micromarketer: see Chapter 5). This positioned CCN in the market as a provider of services through a 'do-it-yourself software' channel by contrast to the 'call in' channel on which users were previously reliant. In response to this market threat CACI started work on the development of its own system, InSite, which was launched in 1988. Pinpoint launched GEOPiN at much the same time. While Mosaic Systems and InSite essentially were geodemographic analysis systems, integrated with mapping written in-house, Pinpoint took a different approach, writing geodemographic applications to run on ESRI PC-ARC/INFO, GIS software.

One reason for Pinpoint's GIS-oriented approach was its chairman, Gurmukh Singh and his interest in such technology. Singh had sat on the Chorley Committee (the Committee of Enquiry for the Handling of Geographic Information) during 1986/7. The Chorley Committee reported in mid-1987 (Department of the Environment, 1987) and raised the profile of GIS in both the commercial and academic sectors (eventually leading to the establishment of the Association for Geographic

Information: www.agi.org.uk). One of the recommendations of the Chorley Report was the use of the unit of address as a geographic 'building block' for GIS-based analysis. Pinpoint's PAC (Pinpoint Address Code) fulfilled this function.

In more recent times Ordnance Survey (UK) has produced Address-Point with normally a 0.1 m resolution coordinate ascribed to residential, business and public postal addresses in Great Britain. (Address-Point is also used to calculate the geographical centre of postcodes recorded in the Code-Point file.) Ordnance Survey has also launched MasterMap, promoted as 'the definitive map of Great Britain'. There are similarities in the development of this product and the creation of a National Spatial Data Infrastructure (NSDI) in the USA (National Research Council, 1993, 1994, 1995; Executive Order 12906 (Clinton, 1994); Federal Geographic Data Committee, 1997; Executive Order 13286 (Bush, 2003)). In fact, recognition of the strategic importance of geographic information for so many areas of governance, business and commerce has led to the development of NSDIs throughout the world. Longley *et al.* (2001, p. 416) list 28 countries that had NSDIs in the year 2000.

Returning to the turn of the 1990s, Infolink joined the geodemographics market by launching DEFINE, which was unusual in having two interlinked classifications: one census based; the other credit based. The Unit for Retail Planning Information (URPI) formed The Data Consultancy and contributed to the continuing integration of geodemographics and GIS by writing its Illumine analysis software to run on MapInfo GIS. GeoMatrix launched its Prospex GIS software, which was subsequently acquired by Beacon Dodsworth. Meanwhile, the University of Leeds, School of Geography, which had been operating in the commercial sector through ULIS (University of Leeds Industrial Services) formed GMAP as an outlet for its gravity modelling expertise in the retail and financial service sectors (Birkin *et al.*, 1996; Birkin, Clarke and Clarke, 2002).

A new census decade started with the release of the 1991 UK Census data in the summer of 1993. Twelve organizations signed up as census agencies (CACI, CCN, CDMS, Capscan, Chadwyck-Healey, Claymore Services, Equifax Europe (UK), GMAP, Infolink Decision Services, Pinpoint Analysis, SPA marketing systems and The Data Consultancy) but only six elected to produce neighbourhood classifications, largely because the cost of (effectively) buying out the census offices' royalty to do so was too expensive. For example, commercial users of the 1991 Census area statistics wanting to have a comprehensive set of variables nationally at the output area level could pay £100,000 – 90% of which was royalties.

Pinpoint was first to market post-1991 geodemographics with the launch of its new PiN and FiNPiN classifications (September 1993). However, as an early sign of the consolidation of the geodemographic industry that was to follow, Pinpoint was taken over by CACI within weeks. CACI's own updated ACORN and CCN's new Mosaic were next to emerge; then CDMS with SuperProfiles and Infolink with DEFINE. EuroDirect launched Neighbours and Prospects. Later, in 1994, Infolink launched PORTRAIT, based mainly on lifestyle data (data about individuals and/or their households, collected from survey information: see Chapter 9). One month later, Infolink was acquired by Equifax; so Equifax instantly had a stake in the geodemographics market.

Outside the market for neighbourhood classifications – but in the associated lifestyles arena – two of the big players (NDL and CMT) came together to form the Calyx Group. This new company took a different view regarding the use of NDL data within PORTRAIT and negotiated its return to the Calyx Group. Calyx was then acquired by VNU, who owned Claritas in the USA and, in January 1997, Calyx was renamed Claritas UK. The Calyx Group subsidiary had already launched a desktop analysis and mapping system (Catalyst) on a MapInfo platform, and also a UK version of PRIZM, built from lifestyle data (therefore, effectively a successor to PORTRAIT). Two more organizations joined the census agency line-up: GEOPLAN and Business Geographics. However, both Pinpoint and Infolink had 'disappeared' through acquisition, and Claymore had relinquished census agency status, so the total number of agencies shrank to 11. CCN changed its name to Experian in June 1997.

Further market developments followed in 1998. Littlewoods decided that CDMS was 'non-core' and closed it down. Claritas UK bought the SuperProfiles classification and signed up as a census agency. GUS, owners of Experian, acquired Metromail in the USA and with it also acquired ICD in the UK (ICD were the third of the 'big three' UK lifestyle database operators, alongside CMT and NDL). ICD was 'folded' into Experian, which at a stroke became one of the two largest UK lifestyle database operators, as well as one of the big two geodemographic agencies – alongside CACI. MapInfo acquired The Data Consultancy and its census agency status. EuroDirect was majority acquired by the Skipton Building Society and in 2000 EuroDirect acquired the MicroVision and DEFINE classifications from Equifax (although Equifax continued to sell what were now EuroDirect's classifications under licence). EuroDirect rebranded its Neighbours and Prospects classification as CAMEO.

In 2001, Claritas and MapInfo announced a strategic alliance across Europe, by which Claritas would supply its data into a joint product (branded TargetPro), but Claritas would adopt MapInfo software as its delivery platform. GMAP was acquired by the Skipton Building Society and put together with EuroDirect within the Skipton Information Group (GMAP being rebranded GMAP Consulting). Claritas Europe was acquired by Acxiom in December 2003 who also subsequently acquired Consodata, which has largely operated in the European consumer data market (Lawson, 2004).

At the end of 2001, a High Court decision had the effect of 'freezing' access to the UK electoral register until 2002 and restricted access thereafter (see Chapter 9, Section 9.4). While this did not affect the use of the electoral register for credit referencing – i.e. to help credit lenders check names, addresses and the potential credit risk of customers – it did create problems for direct marketing and geodemographic applications as some voters chose not to permit their personal information to be passed on to third parties (they 'opted out'). In December 2002 the opt-out rate was just above 20% but rising: the newer, 'rolling register' has an opt-out of nearly 24%; and the December 2003 Register has 26%. The coverage is uneven geographically, with some local authorities in effect encouraging opt-out. In 2003 Bridgend (Wales) had an opt-out of 85%! (sources: Database Marketing, 2003; www.dmarket.co.uk).

3.2 Geodemographics today

Despite the increased use of lifestyle data, the core of most geodemographic classifications is census-based. It follows that there are periodic waves of increased activity within the geodemographic industry and that these reach the commercial shoreline following each new arrival of small-area, census data.

A major change in attitude within the UK has been a move to an American model that recognizes the public utility of making census data as widely available as possible. Accordingly, the census outputs are effectively free at the point of use, and licences for re-use and publication are also free (www.statistics.gov.uk/census2001). The data for England and Wales can be accessed via the Neighbourhood Statistics website (http://neighbourhood.statistics.gov.uk; for Scotland see www.scotland.gov.uk/stats and for Northern Ireland see www.ninis.nisra.gov.uk), which is similar in function to the US Census Bureau website (www.census.gov) – although without the latter's ability to output data in a 'GIS friendly' format (specifically, in the TIGER/Line format: see, for example, Wise, 2002).

However, for academic users in the UK the output of census data in a GIS format has been possible for some time, by means of the CasWeb system (www.census.ac.uk/casweb). GIS vendors such as ESRI (through their ProCENSUS system) have also developed packages to integrate census information with mapping and analytical functionality. MapInfo has implemented a similar scenario with its TargetPro system.

Post-2001, not only are the UK Census data free to end-users but also intermediaries can resell either 'raw' data or derived data such as neighbourhood classifications, without royalty to the various census offices (including the Office for National Statistics, the General Register Office for Scotland and the Northern Ireland Statistics and Research Agency). This situation has made it possible for smaller organizations to enter the industry. Consequently it is anticipated that many more neighbourhood classifications will appear on the market and it will be interesting to see how these new entrants compete (perhaps on a price platform with the established companies trading on their past experience, expertise and customer support: see chapter conclusion for an elaboration on this).

By early 2004, the UK market had seen two new or, rather, updated neighbourhood classification launches: ACORN from CACI and Mosaic from Experian. A further 12 were known about but not yet launched – the main reason being delays to the release of the 2001 Census Area Statistics. Table 3.1 lists the companies who had either launched or declared their intention to launch at the time of writing.

Table 3.1 Neighbourhood classifications announced in the UK post 2001 Census

Company	Classification
CACI	ACORN
GeoBusiness	ATOMICube
EuroDirect	CAMEO
AFD	Censation
Allegran	Gnuggets
Streetwise	Likewise
Business Geographics	Locale
Experian	Mosaic
Beacon Dodsworth	P^2 People and Places
Claritas	PRIZM
ISL	RESIDATA Lifetypes
ONS	2001 Area Classifications
The Clockworks	SONAR
Acxiom	Personicx Geo

Source: After Sleight (2004a, p. 38).

On the face of it, the new version of ACORN looks similar to its predecessor. The 2001-based version has a hierarchy of five top-level categories, which divide into 17 groups and 56 types. The 1991-based version had six, 17 and 56, respectively. However, this apparent commonality masks some very different data inputs and a different build methodology from previous ACORNs. From 1979 up until 2000, ACORN had always exclusively been built from census data. A year 2000 update had introduced lifestyle date for the first time and also signalled a move to a unit postcode classification. Now the 2003 version is built using 2001 Census data, with further information including income, house prices, shareholdings, lifestyle data, electoral register information, the PAF and neighbourhood statistics – the latter, introduced above, being governmental initiatives to compare or download statistics for local areas on a wide range of subjects including population, crime, health and housing.

CACI adopted a new, two-stage classification process that involved selecting census variables to profile neighbourhoods and then clustering those profiles to produce a stage 1 solution. Stage 2 involved classifying postcodes. Starting with the ACORN code of the census OA from stage 1, new datasets were introduced at the more detailed, unit post-code level. Postcodes with data suggesting they did not match their OA classification were reassigned to other neighbourhood types using decision algorithms. The same process will facilitate subsequent updates. The fact that postcodes need to be reassigned relates and to some extent addresses the problem of the ecological fallacy described in Chapter 2. Specifically, it highlights that there can be important socio-economic and other population diversity within apparently uniform neighbourhood units. In this particular case classification at the postcode level helps reveal diversity that a coarser, census classification would otherwise conceal.

Experian's build methodology is discussed in detail in Chapter 6. First, data for all residents and households in the country are collected then combined with data from higher levels of geography (including unit postcodes and OAs). All the input variables are then tested for discrimination, robustness and correlation to other variables. Once the final list of input variables has been selected, a set of input weights is applied as part of the clustering process. The result is a list of variables that have differing importance to the clustering methodology, determined by how well they discriminate at differing levels of geography.

The actual structure of the finished Mosaic classification is similar to the previous version: a two-tier classification, with 11 groups as before but this time there are 61 types, versus 52 formerly. The inclusion of Northern Ireland makes the new Mosaic a full UK version and this accounts, in part, for the larger number of types. A significant innovation is that, on this release, Mosaic classifications can be applied to individual households either on the basis of the attributes of their household or their postcode. The post-code classification is built in a conventional way using data values from a variety of sources and a variety of levels of geographic resolution, such as the electoral register, the postal address file, lifestyle data and the census. In the household-level classification, addresses are matched to best fit Mosaic clusters using information some of which, such as the census, is accessed at OA level, other variables, such as lifestyle data, at postcode level, but a number of key variables, such as those from the electoral register and the PAF at the address level. While all addresses within a postcode will share common values across, for example, the census, different addresses will have different values for variables drawn from address level data. As a result there will be some instances where different addresses within the same postcode will be best described by different Mosaic codes and instances where the best-fit Mosaic code for the address will be different from the best-fit Mosaic code for the postcode in which the address belongs.

Splitting the other, proposed classifications into two groups – and dealing first with the established 'players' in the geodemographics market – new classifications were expected from Claritas and EuroDirect. Claritas are not anticipated to launch an updated version of PRIZM until the third quarter of 2004. It is expected to be built mainly from Claritas lifestyle data, with some census data influence.

EuroDirect have also moved from census-only classifications to ones that include a wider range of data. In addition to 2001 Census data, EuroDirect's latest CAMEO postcode-level classification utilizes: house value data (specifically, council tax bandings weighted by Land Registry data); shareholder data from share registers (the *Financial Times* top 500 companies, i.e. the FTSE 500); information about company directors; credit risk data; information about household composition and length of residency from the electoral register; and building type and age of resident data from EuroDirect's Data Exchange data cooperative. The latest version of CAMEO was launched in March 2004, with 10 groups and 58 types; while previous versions have operated at the scale of census output zones, the new version operates at postcode level. The new CAMEO group codes, descriptions and their share of UK households may be viewed at www.eurodirect.co.uk.

In addition to their area-based classifications, many of the estab-lished geodemographic companies also have a portfolio of individual and household-level classifications, and/or other national 'data lists', intended for a range of direct or interactive marketing applications. Many of these are summarized in Table 3.2. Some of the advantages and disadvantages of these micro-level datasets vis-à-vis the neighbourhood-based approach are considered in Chapter 9.

The second group of post 2001 Census classification authors are essentially newcomers to the market (although some of the developers had previous involvement in building classifications for other companies). Some of the expected characteristics of these new products are reviewed in Table 3.3. Among these are classifications produced by the Office for National Statistics (ONS) at four levels of geography: local authority, ward, health authority and output area. Although ONS has developed classifications of England, Scotland and Wales after each census since 1971, the post 2001 classifications are the first to include Northern Ireland and also the first to extend classification (in conjunction with the University of Leeds) to the most geographically detailed census units (i.e. output areas). These classifications are most likely to be used extensively in public-sector applications, although they could also freely be used in commercial applications, subject to ONS' terms

Table 3.2 Some of the individual/household classifications and sources of lifestyles data available in the UK as of early 2004

Company	Product	Type	Market served
CACI	LifestylesUK	List	General
	PeopleUK	Classification	General
	etypes	Classification	Online
	ChannelChoice	Classification	Channels
	Fresco	Classification	Financial
Claritas	Lifestyle Universe	List	General
	P$YCLE	Classification	Financial
	PRIZM Household	Classification	General
	Onliners	Classification	Online
	ConsumerValues	Classification	General
Consodata	OmniLifestyle	List	General
Equifax	Dimensions	List	General
	MicroMatch	Classification	Financial
EuroDirect	CAMEO Choices	List	General
	CAMEO Lifestyles	Classification	General
Experian	Pixel	Classification	General
	Canvasse	List	General
	Financial Strategy Segments	Classification	Financial
	Touchpoint Segments	Classification	Channels
	Fashion Segments	Classification	Fashion
	Mosaic Household	Classification	General

Table 3.3 New geodemographic classifications expected in the UK following the 2001 Census

Vendor	Product	Data sources used
AFD Software Ltd	Censation 52 residential clusters ranked by affluence (plus other non-residential clusters)	2001 Census Electoral Register Lifestyles data
Allegran Ltd	Gnuggets A segmentation of online behaviour	2001 Census Website registration data (e.g. age, sex, address) Advertising 'click through' data Online purchasing data
Beacon Dodsworth	P^2 People and Places Being built under the guidance of some of the developers of 1991 SuperProfiles	2001 Census PAF Retail access data Neighbourhood Statistics
The Clockworks	SONAR Two-tier hierarchical classification: 80 detailed clusters; 24 groups (characterized by lifestage and relative wealth)	2001 Census A postcode-based wealth classification Consumer activity data (e.g. new car ownership, credit card usage, house-moving) 2002 Land Registry house price data
GeoBusiness Solutions Ltd	ATOMICube Will be available at three levels of solution at differing prices	2001 Census Non-Census data (e.g. income and household data)
Intermediary Systems Ltd (ISL)	RESIDATA Lifetypes A market-specific product used by the insurance industry. Three-tier classification: 100 clusters aggregate into 35 segments and then 10 groups	2001 Census RESIDATA housetypes classification PAF Spatial variables (e.g. distance to major shopping centre, distance to motorway junction)
Office of National Statistics (ONS)	2001 Area Classifications The local authority classification, for example, has been produced at three hierarchical levels: there are eight clusters at supergroup level, 13 clusters at group level and 24 clusters at subgroup level	2001 Census At local authority, health area and ward level scales; and for output areas (forthcoming, with the University of Leeds)
Streetwise Analytics Ltd	Likewise Expected to be an address-level, postcode and possibly OA system	2001 Census House price data (to help differentiate income from cost of living)

and conditions. Further details are at www.statistics.gov.uk (under National Statistics 2001 Area Classification).

Case study: the US market for geodemographics

Dave Miller, senior vice president, Data Research and Development, Claritas USA

The release of the statistics from successive US Censuses increasingly has been an important business event over the last 30 years. The impact started with the first release of low-level, machine-readable census information, at the postal area level, in the mid-1970s. The relative ease of joining this with consumer information spawned the first US geodemographic segmentation systems and jump-started the demographics industry.

When looking into this industry in the USA, one of the principal considerations to be kept in mind is that, in contrast to the UK or Canada for example, the US Census has not been copyrighted. As a result, the information has always been 'freely' available for redistribution. This does not mean that acquiring the data is completely free! There has always been a stipulation that the cost of producing the media was recoverable. In the 1980s this meant that a reel of tape would cost about $150. As the entire census at low level required many hundred tapes, this cost did add up. In response, the US industry has competed on the value-added aspects in terms of simplifying organization of the data, ease of access, simplicity of use, customization to business sector and the creation of inter-censal updates.

The release of a census has always provided a good opportunity for a start-up business that focuses on the repackaging and redistribution of the census data. Although this opportunity has declined somewhat with the improved access provided by the Bureau of the Census itself, there is a wide array of companies providing access to the Census 2000 data. However, the business niche of maintaining and updating the census information is quite small. There are only five companies providing regular inter-censal updates in the USA at a low geographic level (e.g. block group). They are: Claritas; Easy Analytic Software Inc.; ESRI Business Information Solutions (ESRI-BIS, newly incorporating the demographic arm of CACI); MapInfo; and Applied Geographic Solutions (AGS). Of these companies, those with the longest history of providing data are Claritas and ESRI-BIS both of which can trace their roots to the start of the geodemographic business in the mid-1970s. The users of this information can be found in every business sector and virtually all major companies obtain and use this type of data.

Although the production of demographic estimates and projections is one aspect of the geodemographic industry, the integration of this information into software that addresses various business needs is an equally important sector that is also invigorated by the release of new census data. The players in the site location and evaluation business are: Claritas; MapInfo/Thompson; SRC; ESRI-BIS; AGS; GeoVue; and ScanUS. This market is vibrant and very competitive and the changes in the product lines are continuous. Direct changes in the market due to the census are somewhat small.

The release of the census information does, however, have a large impact on the design of the geodemographic segmentation systems. The businesses space for geodemographic oriented segmentation systems includes: Claritas (PRIZM, P$YCLE and ConneXions); ESRI-BIS (ACORN and Community Tapestry); MapInfo (PSYTE); and Experian (Mosaic). As this is written, all are involved in updating their segments.

The need for updating segments reflects a number of demographic shifts that characterized the 1990s in the USA. The ageing of the baby-boom generation has moved this important and large segment from early mid-life to late mid-life with the associated transition from family-raising to child-launching. A second aspect has been the loss of manufacturing jobs and their replacement with service sector jobs. The greying of America not only reflects the shift in age but also the movement from traditional 'blue-collar' manufacturing to 'grey-collar' service. Further, recent data from the 2000 Census shows that the trend in increasing education levels has for the first time resulted in more than half of the adult population having had some exposure to college. The cumulative effect of these and other changes is both to alter the way we measure the population and socio-economic characteristics that are important in segmentation, and to modify the balance of segment sizes. It is useful to note that while the underlying concepts behind geodemographic segmentation systems (affluence, family composition, age, etc.) have not changed over the years, their census indicators and balance is continually shifting. It is this effect that leads to the need to redevelop the classification segments: to reflect changing populations and changing geographies of population.

At this point most of the systems are either in the early stages of deployment or in the late stages of development. Claritas made a number of changes in its PRIZM system and extended it to the household level providing both geodemographic assignments (ZIP+4 and others) and household-level assignments in the same schema. ESRI-BIS redeveloped its ACORN system as Community Tapestry. The system is planned to provide segment assignments at the ZIP+4 level. An overview of the US Census and postal geographies is provided by Table 3.4. It can be seen there that ZIP+4 is a close approximation to the UK unit postcode, although a little less populated: the average unit postcode in the UK contains about 12–16 households.

MapInfo's PSYTE system was updated at the census block group level but has maintained the same schema as in 1990. Experian's new Mosaic was launched in the fall of 2003, with 12 groups and 60 types. It operates at ZIP+4 level, with a household-level equivalent due in the third quarter of 2004. A demographic segmentation system but one that does not provide geographical focus is Personicx from Acxiom. This is a demographic, household-level segmentation system developed from characteristics maintained on the Acxiom compiled (lifestyle) list.

The segmentation space is also shared by a number of other non-geodemographic systems, including: Cohorts from Looking Glass; VALS from SRI Consulting Business Intelligence; and MindBase from Yankelovich. The Cohorts system was developed from survey data and incorporates behaviour, demographics and lifestyles. VALS from SRI is one of the earliest attitudinal-oriented systems. VALS provides differentiation based on survey data. MindBase from Yankelovich couples certain demographics with survey-based data to create its attitudinal segments. One focus of these systems is on the creation of different advertising messages to each segment. Attitudinal segmentation is not, however, new: a 'socio-styles' system was developed by French researchers during the 1970s (Cathelat, 1990; Birkin, 1995).

Table 3.4 Common US geographic levels

Level	Areas	Average number of households
Census		
State	51	2068237
County	3141	33582
Tract	65443	1612
Block	208790	505
Group		
Block	8205582	12.9
*Postal**		
ZIP Code	29800	3540
ZIP+4	29000000	3.6

* Please note that the postal counts represent residential codes and do not include business or other types

3.3 The role of market research linkages

Commercial uses of geodemographics are much enhanced by the ability to link them to market research. As we noted earlier, the precedent for this was set in the late 1970s, when BMRB executives tested the electoral ward, CRN classification on their Target Group Index (TGI) survey. Following this initiative, the TGI played an important role in the developing the UK geodemographics market – and it still does today! The reason for this is that BMRB has encouraged companies that develop and market commercial neighbourhood classifications to provide 'look-up' tables of geodemographic codes for unit postcodes, thus permitting the survey respondents to be coded by neighbourhood type. This enables the survey results to be grouped by neighbourhood and so the users of the products, brands and media covered by the TGI survey can be profiled. An example is provided by Table 3.5.

The geodemographic coding of market research helps inform understanding of products, brands and media in terms of the type of people who consume them, with what frequency, with what 'weight of consumption' and with what level of monthly value. Moreover, the information can directly be used in a key application of geodemographics – small-area market estimation. By deriving a consumption profile from a source such as the TGI, it is then possible to relate the profile to other small areas of the same neighbourhood type in the UK and thus estimate the likely market value of a particular product in a particular area or region. The methodology is explained by one of our contributors in a study used to validate geodemographics (Chapter 8, Section 8.3; see also the case study on using geodemographics in the restaurant market: Chapter 1). As a caveat, the estimation is improved by constraining national totals to some reliable known source (such as national income and expenditure data) and by taking into consideration regional and other important geographical differences.

3.4 Use of non-census data

During our account of the history of commercial geodemographics, reference has been made to the growing use of non-census data in neighbourhood classifications. Such data have included: credit references and county court judgements; postal address files; electoral registers; share registers; company directors information; unemployment data; insurance

Table 3.5 Target Group Index cross-tabulation by ACORN – households with unit trust investments

	ACORN 2001 groups	Profile (×1000)	%	Base (×1000)	%	Penetration	Standard error	Z-Score	Index
1.A	Wealthy Executives	306	14	1640	6	19	0.76	10.5	225
1.B	Affluent Greys	265	13	1608	6	16	0.72	8.6	198
1.C	Flourishing Families	224	11	1955	8	11	0.67	4.3	138
2.D	Prosperous Professionals	111	5	597	2	19	0.48	6.0	224
2.E	Educated Urbanites	120	6	1012	4	12	0.50	3.4	143
2.F	Aspiring Singles	63	3	984	4	6	0.37	−2.4	77
3.G	Starting Out	45	2	749	3	6	0.31	−2.6	72
3.H	Secure Families	355	17	4323	17	8	0.81	−0.2	99
3.I	Settled Suburbia	162	8	1775	7	9	0.58	1.2	110
3.J	Prudent Pensioners	70	3	591	2	12	0.39	2.5	143
4.K	Asian Communities	13	1	198	1	7	0.17	−1.0	79
4.L	Post-Industrial Families	69	3	1264	5	5	0.39	−4.4	66
4.M	Blue-Collar Roots	114	5	2714	11	4	0.49	−10.7	51
5.N	Struggling Families	152	7	3943	15	4	0.56	−14.8	46
5.O	Burdened Singles	28	1	1289	5	2	0.25	−15.1	26
5.P	High-Rise Hardship	7	0	396	2	2	0.12	−9.8	21
5.Q	Inner City Adversity	14	1	447	2	3	0.18	−6.2	38
6.U	Unclassified	0	0	0	0	0	0.00	0.0	0
	Total	2118	100	25485	100	8			

ratings; retail access information; house price data; and, increasingly popular, lifestyle data. Indeed, lifestyle data have now extensively been used in neighbourhood classification, and also for building individual- and household-level classifications for micromarketing.

Many of these various datasets have been used in imaginative, if not necessarily sophisticated ways! Postal address files, which are available for many counties including Australia and the UK, can be used to shed light on certain aspects of neighbourhoods. For example, the presence of 'Farm' in an address may be indicative of a rural area, particularly if the nearest neighbour properties also appear to be farms. The street name 'Royal York Terrace' is a good clue to the type of buildings that are to be found there. And houses with names rather than numbers may be a useful indicator of status!

Electoral registers usually record the name, date of birth (eligibility to vote) and gender of each member of a household. Such data, where they are made available, can be used to model family composition and, by comparing successive registers, length of residency can be estimated. Fore and surnames are also of interest because they are indicative of age, since certain names come in and out of fashion. One of us shares the 99th most popular first name for baby boys in England and Wales in 2003 (www.statistics.gov.uk) while another of us is not in the top 100, though his son's name is 52nd! Table 3.6 shows that in 2003 the most popular name in the two countries was Jack for baby boys and Emily for baby girls. In 1904 the most popular boys' name was William; the most popular girls' name, Mary.

Table 3.6 The most popular names for babies in England and Wales, various years 1904–2003

		1904	1934	1964	1994	1999	2003
Males	1st	William	John	David	Thomas	Jack	Jack
	2nd	John	Peter	Paul	James	Thomas	Joshua
	3rd	George	William	Andrew	Jack	James	Thomas
	4th	Thomas	Brian	Mark	Daniel	Daniel	James
	5th	Arthur	David	John	Matthew	Joshua	Daniel
Females	1st	Mary	Margaret	Susan	Rebecca	Chloe	Emily
	2nd	Florence	Jean	Julie	Lauren	Emily	Ellie
	3rd	Doris	Mary	Karen	Jessica	Megan	Chloe
	4th	Edith	Joan	Jacqueline	Charlotte	Jessica	Jessica
	5th	Dorothy	Patricia	Deborah	Hannah	Sophie	Sophie

Source: Office for National Statistics.

The first company to exploit these changing fashions was CACI with its MONICA system in the 1980s. The examples given by CACI were clearly associated with age cohorts and included: Michelle, Sharon, Kevin and Gary (then 18 to 24 years, now 40 somethings!); Pamela, Janet, Philip and Brian (24 to 44 years); Sylvia, Brenda, Kenneth and Raymond (45 to 64 years); Hilda, Ethel, Percy and Herbert (then 65+ years, now a rare breed!). There is potential to develop this approach to consider not only age but also regional differences. As we noted in the conclusion to Chapter 2, a project looking at the current and historical geographies of surnames is taking place at the Centre for Advanced Spatial Analysis, University College London.

3.5 Conclusion

In this chapter we have reviewed some aspects of the commercialization of geodemographics, and its growth into an international and multimillion dollar industry. What is striking is how a simple idea has endured, evolved and been 'repackaged' into a number of different products.

We have three suggestions to why this has happened. The first is: *because it is simple*! Not only that, it has proved to be useful in a wide range of applications. These two facts are not unrelated: the simplicity of the methods makes it easy to understand, apply and transfer to a variety of analytical contexts. Furthermore, the basic method can be 'tweaked' to best suit a particular setting – perhaps by customizing the classification with the most relevant sources of data as in market-specific classifications.

The second reason was alluded to by David Miller in the case study: society does not sit still. Many people move and those that don't certainly age! Consequently, with each new release of, in particular, census data there is a need to rebuild classifications to better reflect the demographic, social and economic patterns of the time. If nothing ever changed there would be little need for a census of the population – it would be little more than a number-generating exercise!

Third, the growth of geodemographics has coincided with an increased interest in the utility of geographical information and in methods (both technological and analytical) to make better use of it. The two trends are mutually reinforcing. Building an accurate neighbourhood classification first requires accurate data and the technological platform to handle it. Interest in the classification generates demands for improvements, ideally leading to better data, better methods, better classifications and so forth.

These underlying reasons for market growth have been exploited by the geodemographic suppliers, which have built a thriving and sophisticated market. Over the last two decades, robust business models have been developed which have the following characteristics:

- Suppliers can define key types of analysis that cater for a very high proportion of client business needs and embed these in the analysis and mapping software they supply. As the number of experienced users grows and they become familiar with these tools there is less requirement for consultancy advice.
- Most large users purchase a package of data plus software. The software is customized to handle the vendor's proprietary classifications, plus other bundled data (see Chapter 5). Therefore, rolling contracts are the norm, rather than once-only sales.
- Annual data updates are one reason for annual licensing arrangements. Datasets such as population updates and changes to postal geography must be updated at least annually by users and vendors can help achieve this. Training, helpdesk support and general consultancy advice is also included in the annual contract.
- The established suppliers back their classifications with extensive support material (for example, multimedia visualization of the neighbourhood types: see Chapter 5). They believe this is valuable to support both internal 'experts' within client companies (who may need to pass on understanding to non-experts) and also to give more 'qualitative' understanding of different types of neighbourhood within the vendor's own team.
- In cases where the classification is being used as a long-term strategic tool within a client organization, this level of qualitative understanding is particularly important, given the need for a clear understanding of the target market; the likelihood of needing to disseminate this knowledge widely within non-specialist departments is much greater.

Taken collectively, the business practices employed by the established suppliers represent their brand equity and their best strategy to defend their market shares against newcomers. It is noticeable that new entrants to the market tend to acknowledge this and adopt different strategies – mainly price-based. Only time will tell how much impact the new entrants will have on directing the continued evolution of the industry.

Summary

- Two of the earliest commercially marketed geodemographic classifications were ACORN in the UK and PRIZM in the USA.
- The former built on work at the Centre for Environmental Studies looking at how types of deprivation varied by neighbourhood.
- The development of geographical information-handling technologies has permitted linkages to be made between census, postal and, more recently, individual/household datasets, fuelling the growth of the geodemographic and data-selling industries.
- Although the geodemographic industry is characterized by consolidation and merger, new entrants to the market have been encouraged by free access to census datasets.
- Data vendors increasingly incorporate a mix of census and other data when building and updating classifications.
- Established geodemographic vendors have developed a business model that is characterized by the annual licensing of the classification itself, supporting software and analytical tools, data updates and consultancy services. Together these form their 'brand equity'.

Further Reading

- Charlton, M., Openshaw, S. and Wymer, C. (1985) Some new classifications of census enumeration districts in Britain: a poor man's ACORN, *Journal of Economic and Social Measurement*, **13**, 69–96.
- Sleight, P. (2004) *Targeting Customers: How to Use Geodemographic and Lifestyle Data in Your Business*, World Advertising Research Center, Henley-on-Thames.
- Webber, R. (1985) The use of census-derived classifications in the marketing of consumer products in the United Kingdom, *Journal of Economic and Social Measurement*, **13**, 113–24.
- *Journal of the Market Research Society*, Special Issue on Geodemographics (1989, **31**, 1–150).
- The Geodemographics Knowledge Base: www.geodemographics.org.uk.

4
Geodemographics and GIS

Learning Objectives

In this chapter we will:

- Define geographic information systems (GIS) and consider some of the ways they are used to manage and make sense of geographic information.

- Introduce some of the principles of GIS, and some of the methods and ideas of the interdisciplinary field of geographical information science.

- Consider the ways in which GIS and geodemographics may be regarded as complimentary technologies.

- Focus on the vector data model in GIS and how it represents real-world entities as points, lines or polygons.

- Look at some methods for mapping geodemographic information with GIS and some of the problems of doing so.

- Offer a commercial perspective on using GIS for neighbourhood analysis and targeting, written by Stewart Berry of Caliper Corporation.

Geodemographics, GIS and Neighbourhood Targeting   Richard Harris, Peter Sleight and Richard Webber
© 2005 John Wiley & Sons, Ltd   ISBNs: 0-470-86413-3 (HB); 0-470-86414-1 (PB)

Introduction

Geographic information systems have been described as a set of technologies that help us to see our small blue planet in better ways (Longley *et al.*, 1999). More commonly referred to by the acronym GIS, applications include: local governance; business and service planning; logistics; and environmental management and modelling (see Longley *et al.*, 1999, Part 4; Longley *et al.*, 2001, Chapter 2). In both public- and private-sector research, GIS are used to manage geographic information, help identify geographical trends and patterns and to model spatial processes.

However, GIS have been described as a 'nearly' technology for marketers (McLuhan, 2003). Beyond the hype, the actual use of GIS presently is limited to the larger retailers and suppliers, with little expansion into marketing applications. This, despite widespread agreement that the true value of geographical information is only revealed once that information is analysed geographically! McLuhan (2003) cites a survey by GeoBusiness Solutions revealing that only 28% of company boards fully understand the operation and marketing benefits of GIS, with the perceived (and often, actual) high cost of investing in GI software and data products being one of the barriers to GIS reaching its potential.

In this and the following chapter we agree with those who see virtue in GIS as tools for managing, analysing and visualizing geographical information; who use them to look at where things happen (or not), to help explain what occurrences are associated with which places and why, and to use that knowledge for planning and management. Yet, we shall also argue that the sorts of tools and methods required by a 'typical' geodemographic user are not necessarily those provided by a standard desktop GIS, which has its origins, for instance, in environmental research. A distinction will be made between geographical information systems (GIS) in a broad sense and geodemographic information systems (GDIS) that cater for a more specific clientele.

The differences between GIS and GDIS will be explored fully in Chapter 5. In this chapter we set the scene for that discussion by first outlining some of the more general principles of neighbourhood analysis that are conducted within a computerized environment and by providing an introduction to some of the principles, theories and methods of the interdisciplinary subject known as geographical information science (GISc). Readers who wish to 'dig deeper' into this field should refer to the suggested further reading listed at the close of this chapter.

4.1 Principles of GIS

With the proliferation and diversity of GIS applications it is not surprising that there are many definitions of GIS, each reflecting the nuance and perspectives of different users' interests. Nevertheless, there is much shared ground and this leads to a commonly recognized view of GIS as a system of 'component tools used to capture, store, transform, analyse, and display geographical data' (Haggett, 2001, p. 719). Such tools are integrated into the off-the-shelf, desktop GIS software that include ESRI's ArcGIS (www.esri.com), Autodesk Map (www.autodesk.com), Intergraph's GeoMedia Professional (www.ingr.com), MapInfo (www.mapinfo.com), Clark Lab's Idrisi (www.clarklabs.org) and Caliper's Maptitude (www.caliper.com).

In addition to these commercial software, there are a number of open source and free GIS available (see http://freegis.org/), of which probably the most well known is GRASS (http://grass.baylor.edu/). There are also viewers available to open and manipulate maps created in GIS packages, including ERMC's MapWindow (www.mapwindow.com) and ESRI's ArcExplorer (www.esri.com).

Haggett's definition of GIS raises the question of 'what are geographical data?' Literally these are data that describe processes, events or activities that take place on or near the Earth's surface and which record where those process, events or activities take place. The key characteristic of a geographic dataset is that it contains both *attribute* information and information about *location*. Attribute information is about the process, event or activity being measured (e.g. the temperature is 26 °C; the building density is 30 dwellings per hectare; the customers on average visit once a week). Location adds where the data were collected or apply to (e.g. the temperature in Paris is 26 °C; the building density within postcode BS16 7DX is 30 dwellings per hectare; the customers visiting the Southbridge store at grid reference $x = 529600$, $y = 181500$ do so on average once a week).

There are several terms to describe the act of assigning location to attribute information, including georeferencing (our preference), geolocating and geocoding (Longley *et al.*, 2001). It is the act of georeferencing that transforms data from being non-geographical (aspatial) to geographical (spatial). Georeferencing data is usually a first step towards the mapping and analysis of attribute information within a GIS to create, for example, maps of customer records, population data or neighbourhood statistics.

Table 4.1 records earnings data from the 2001 Census for Canada, its provinces and territories. The first column of the table (labelled

Table 4.1 Selected earnings data from the 2001 Census for Canada

Name	Total	Earnings groups			Average earnings ($, 2000)	% change, average earnings (1990–2000)
		Less than $20,000	$20,000–$59,999	$60,000 and over		
Canada	16,415,785	6,659,395	7,728,475	2,027,915	31,757	7.3
Newfoundland and Labrador	251,545	133,790	100,270	17,485	24,165	9.8
Prince Edward Island	77,750	42,690	31,225	3,835	22,303	3.5
Nova Scotia	468,825	222,970	209,635	36,220	26,632	4.1
New Brunswick	388,855	193,325	168,495	27,025	24,971	3.3
Quebec	3,815,265	1,583,180	1,860,375	371,715	29,385	3.0
Ontario	6,319,530	2,334,525	3,016,460	968,545	35,185	9.3
Manitoba	609,575	274,705	286,810	48,055	27,178	5.1
Saskatchewan	534,350	263,560	228,345	42,455	25,691	6.3
Alberta	1,768,435	723,715	802,980	241,745	32,603	11.5
British Columbia	2,128,550	865,860	1,001,470	261,220	31,544	4.6
Yukon Territory	18,780	7,255	8,770	2,755	31,526	0.4
Northwest Territories	21,955	7,610	9,410	4,940	36,645	1.4
Nunavut	12,355	6,215	4,225	1,915	28,215	7.1

Source: Statistics Canada's Internet Site, http://www12.statcan.ca/english/census01/products/highlight/Earnings/, 16 May 2003.

'Name') contains the location component: the name of the territory or province. The remaining columns contain attribute information: the distribution of each region's population across the earnings groups; average earnings; and percentage change (1990–2000). Table 4.2 gives the total population for the same geographical divisions as those reported in Table 4.1. Again, the first column gives location and the remaining columns give attribute information.

It is easy to combine the information shown in Tables 4.1 and 4.2 and calculate, for instance, that an estimated 54.7% of the Canadian population were economically active and earning at the time of the census (= 16,415,785 ÷ 30,007,094 × 100). Regionally, the economically active proportion of the regional population is highest in Yukon Territory (18,780 ÷ 28,674 = 65.5%) and lowest in Nunavut (12,355 ÷ 26,745 = 46.2%).

It is possible to make these calculations because both tables contain the same georeferencing system: the name of the territory or province. The common georeferencing permits a relational join to be made between the two tables. The join bridges the two tables in the way Figure 4.1 illustrates. Figure 4.2 shows the same relational join but this time expressed as a Query within Microsoft Access.

Table 4.2 Selected population data from the 2001 Census for Canada

Names	Population		% change	Total private dwellings, 2001
	2001	1996		
Canada	30,007,094	28,846,761	4.0	12,548,588
Newfoundland and Labrador	512,930	551,792	−7.0	227,570
Prince Edward Island	135,294	134,557	0.5	55,992
Nova Scotia	908,007	909,282	−0.1	403,819
New Brunswick	729,498	738,133	−1.2	313,609
Quebec	7,237,479	7,138,795	1.4	3,230,196
Ontario	11,410,046	10,753,573	6.1	4,556,240
Manitoba	1,119,583	1,113,898	0.5	477,085
Saskatchewan	978,933	990,237	−1.1	431,628
Alberta	2,974,807	2,696,826	10.3	1,171,841
British Columbia	3,907,738	3,724,500	4.9	1,643,969
Yukon Territory	28,674	30,766	−6.8	13,793
Northwest Territories	37,360	39,672	−5.8	14,669
Nunavut	26,745	24,730	8.1	8,177

Source: Statistics Canada's Internet Site, http://www12.statcan.ca/english/census01/products/ standard/popdwell/, 16 May 2003.

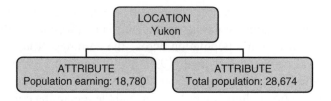

Figure 4.1 Relating tables by a common georeference

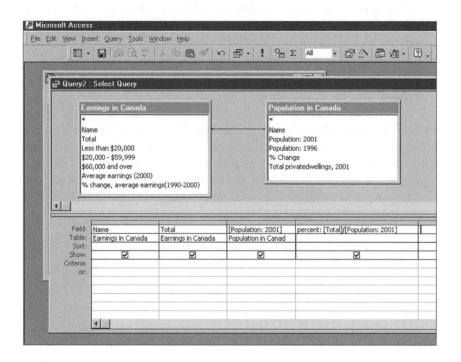

Figure 4.2 Using Microsoft Access to achieve the relational join of Figure 4.1

Consider again Tables 4.1 and 4.2. Both of these contain a set of georeferences and a set of attribute data. It is geographical information, yet at this stage it is not easily mapped in a GIS because to the computer a name like Newfoundland and Labrador is simply bits of ASCII text. They have no intrinsic geographic meaning that can be comprehended by the computer, in the same way that one of us revealing our address to be 15, Camberwick Green is unlikely to mean much to you. It is not then sufficient to have the names Newfoundland and Labrador only. What also is required is one or more geographical coordinates that define where the province is located in geographical space and which can be encoded and stored in a machine-readable database.

A conceptually simple way of defining the locations of Newfoundland and Labrador is to consider their perimeter (their outer edge or border). In the same way that the perimeter of a square consists of four lines (one for each of the four sides) and if we walk around the perimeter we will eventually get back to where we started, so the perimeters of Newfoundland and Labrador can each be represented by a series of line segments that link together to form a closed, two-dimensional shape – a polygon. Quite how good a representation of the perimeter each shape will be depends on how many segments there are relative to how complex a shape the real-world object has. As their number decreases, then so too does the detail of the representation and the outline is said to be more generalized. It is not correct to presume that generalization ought therefore be avoided, since the greater the detail the greater the storage and processing demands on the computer. A more parsimonious (simple) representation may be preferred if it remains fit for purpose, consistent with the detail offered by other sources of information or if the additional expense of increasing the complexity produces only marginal analytical gains. (We saw in Chapter 1 that a similar 'defence' can be made of the 'elegant simplicity' of geodemographic classification.)

A polygon can be defined by the line segments that form its perimeter. In turn, a straight-line segment may be defined by a start point and an end point; or, more correctly, by a start node and an end node. Nodes are portrayed by the map symbols $\bigcirc$ and $\diamondsuit$ in Figure 4.3(a). The location of each node is defined by a grid reference. For example, the enlarged and shaded node is at position (0.81, 1.33). Linking the nodes together in the manner of a 'join the dots' puzzle produces an outline map of the province of Newfoundland and Labrador – Figure 4.3(b). What we have created is a boundary file of the province. The boundary file utilizes the vector data model that is common in GIS research and analysis. For a larger study region involving more provinces and territories the approach can be extended to incorporate further information such as recording the province to the left or right of each line segment. To do so encodes what is known as topological information in the geographical database and is useful for increasing the speed of various analytical operations, error checking and preventing the unnecessary duplication of the same line segment shared by two or more neighbouring polygons (Wise, 2002).

The vector model typically represents real-world objects and entities as points (zero dimension), lines (one-dimensional) or areas/polygons

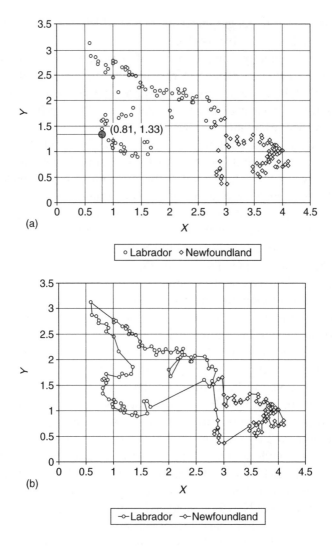

Figure 4.3 Defining the boundaries of Newfoundland and Labrador. (a) Nodes define the start and end of line segments. (b) The line segments define the boundaries

(two-dimensional). These characterize the dimensionality of many of the more common objects represented in a GIS. In a drive-time analysis to define the catchment area of a retail store, the store's location could be defined as a point and the surrounding road network as a series of line segments. Assuming that customers are unlikely to travel for more than 30 minutes to reach the store permits the outer boundary of the store's catchment to be estimated, given knowledge of the length of the roads and the speed of traffic movement along them. The store's catchment area can then be represented as a discrete polygon, also known as a buffer zone around the central point. Linking the catchment model to a geodemographic classification of neighbourhoods permits analysis of the types of neighbourhood and, by extension, the types of consumer to be found in proximity to the store – Figure 4.4; see also Chapter 5, Section 5.2.5.

A boundary file will be required in most applications of mapping neighbourhood data. They are usually obtained from national statistics agencies such as the US Census Bureau (www.census.gov), UK National Statistics (www.statistics.gov.uk), Statistics Canada (www.statcan.ca) or the Australian Bureau of Statistics (www.abs.gov.au); or from an online data warehouse such as ESRI's Geography Network (www.geographynetwork. com). Once a suitable boundary file has been obtained, then provided it can be linked to attribute information through common georeferencing, the attribute data can be visualized. Figure 4.5 maps some of the income information given in Table 4.1.

Of course, dimensions other than zero, one and two do exist: a representation of a city can be more lifelike when modelled with a third dimension (height, giving volume to buildings) (Raper, 2000; Unwin and Fisher, 2001; Hudson-Smith and Evans, 2003); modelling flows such as pedestrian movement along a street requires consideration of a further dimension – time (Batty, 2003); and there is, in fact, no need (conceptually,

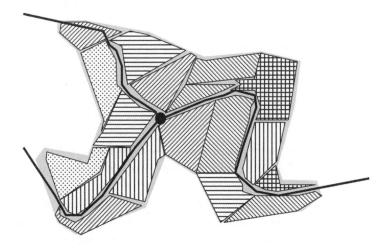

Figure 4.4 Vector representation of the geodemography of a retail store's catchment area. The store is modelled as a point, the roads as line segments and the catchment as a polygon. The model is linked to a classification of the types of neighbourhood within the catchment; the different shadings indicate different neighbourhood types

85

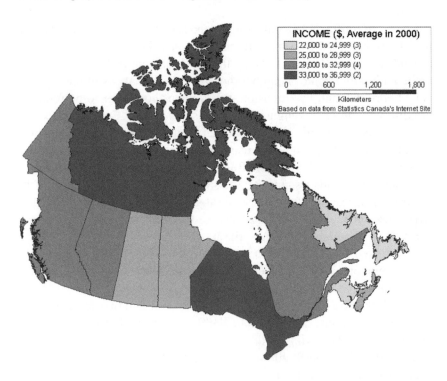

Figure 4.5 Average incomes in Canada. Note: Boundary file based on digital data from the National Atlas Base Map series (http://atlas.gc.ca). Those data are used under agreement but © 1999, Government of Canada with permission from Natural Resources Canada

at least) to restrict dimension to whole numbers: fractions of dimension, or fractals, usefully have been employed, for example to measure decentralization as people move out of cities (a fractal dimension of 1.3 suggests a less compact city than one with 1.7 – the current fractal dimension of London – because less space is filled) (Batty and Longley, 1994; Ravilious, 2004).

Nevertheless, points, lines and polygons are the typical shapes to be seen on the screen of a vector GIS, revealing its origins in cartography. Like traditional maps the vector approach is well suited to the mapping and presentation of much geodemographic information. Some GIS software exclusively is vector based. A second data model to be found in the GIS literature is the raster model, often associated with remote sensing (aerial photography and satellite observation) and environmental process modelling.

An example of a raster GIS is PCRaster (http://pcraster.geog.uu.nl). Raster data can be understood as a grid with values: the raster data model uses a regular grid to cover the study region, each grid cell has a value assigned to it and that value is in some way characteristic of the geographic feature or phenomenon to be found at the location occupied by the cell (see Longley *et al.*, 2001; Chang, 2003). Accordingly, a raster grid can be defined by a simple text file that stores: the geographical coordinates of one corner of the grid; the number of cells in the '*x*' direction (the number of columns); the number of cells in the perpendicular '*y*' direction (the number of rows); the real-world length of a side of the cell square (e.g. 500 metres); a no data value (e.g. -9999) to tell the computer that no reading or observation has been recorded at any cell location assigned that value; and then a series of attribute values – one for each cell. An example of this type of file is given in Chapter 9, Section 9.1.1. The no data value is different from a value of zero. The latter usually records a *known* absence of a particular geographic phenomenon from the cell location. A no data value indicates that whether the phenomenon is present or not is unknown.

Figure 4.6 gives a coarse raster representation of Newfoundland and Labrador. To store this information in the computer a value should be given to each grid cell. One option would be to give cells representing the land mass of Newfoundland a value of one, cells representing Labrador a value of two and all remaining cells a value of zero. This is a simple and sensible way of encoding the model. However, since another user may come along and wonder what it is the values zero, one and two represent, it would be as well to record in an associated 'look-up table' that they indicate sea, Newfoundland and Labrador, respectively.

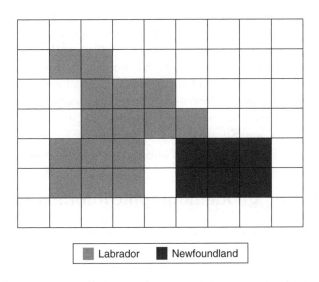

| Labrador | Newfoundland |

Figure 4.6 Raster representation of Newfoundland and Labrador

The blocky nature of Figure 4.6 (albeit exaggerated) reveals why the vector model usually is preferred for mapping demographic and socio-economic information – it is more aesthetically pleasing. An obvious way to reduce the 'blockiness' (pixilation) of the raster image is to increase the number of grid cells which, for a fixed study area, means reducing their individual size. There are two caveats. First, halving the length along any side of the cells will quadruple the total number of cells required. Each of these cells must have a value associated with them and these must be stored somewhere in the computer's memory. Although a number of techniques are available to compress raster datasets and so reduce their storage requirements (such as run length encoding, quadtrees and wavelets, see: Longley *et al.*, 2001; Wise, 2002; O'Sullivan and Unwin, 2003), raster data are often very demanding of memory and produce large file sizes.

Second, it would be misleading but possible to increase the apparent resolution of the cell sizes beyond that afforded by the accuracy and precision of the source data. If the original data were recorded with an accuracy of ± 1 km it would be erroneous to present that same data within a 100 m $\times$ 100 m grid. The same note of caution is applicable to vector representations that can give a false visual impression of the apparently high precision and accuracy of the underlying data.

In practice, the vector data model frequently is used to represent discrete, geographical objects – those with or deemed to have definite borders or edges – while the raster model commonly is associated with continuous fields such as elevation. The distinction between discrete objects and fields raises an important question in the context of neighbourhood classification and analysis: to what degree can neighbourhoods and the populations they contain adequately be represented as discrete (neighbourhood) objects with definite boundaries, as they usually are? We discuss this issue further in Chapter 8. For now it is noted that a number of techniques are available to transform object-based models of population distributions into often more realistic, continuous field views and that these often involve a vector-to-raster data conversion. Some of these techniques are reviewed by Donnay and Unwin (2001) and also at http://census.ac.uk/cdu/software/surpop.

4.2 Mapping geodemographic information with GIS

We have commented that geodemographics is the analysis of people by where they live. Linking this to a GIS framework, the knowledge of *where* people live, work or socialize is the location component.

Data about *what* people do to live, work or socialize is the attribute information. In other words, Location + Attributes = Geography + Demographics = Geodemographics!

Neighbourhood classifications often provide the base map onto which other sorts of geodemographic data are overlaid and analysed. Consider Figure 4.7. This shows a neighbourhood classification of the City of Bristol, England created from the Leeds Geodemographic Analysis System, online at the Centre for Computational Geography at the University of Leeds (www.ccg.leeds.ac.uk). It is a sub-region of a national classification of 1991 UK Census enumeration districts into 10 neighbourhood types (or geodemographic clusters). Only nine of those

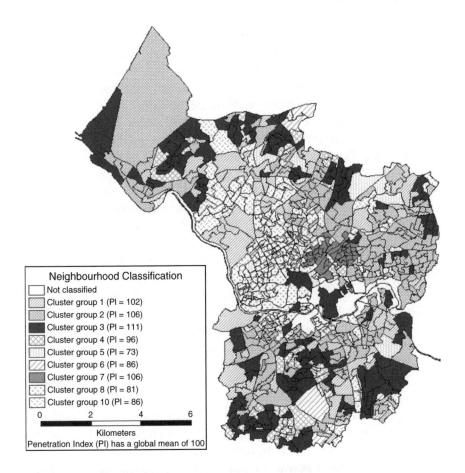

Figure 4.7 Geodemographic classification of Bristol, England, shaded according to the level of readership of *West Country World* within each neighbourhood cluster group. The mean readership across all groups has an index value of 100

89

clusters are shown in Figure 4.7 because the city, unsurprisingly, does not include any of the rural type (the tenth group). Some areas in Bristol either are entirely or predominantly non-residential and are shown as not classified. Although being non-residential does not preclude classification, it does require the use of data drawn from somewhere other than a national census of (residential) population. In most geodemographic classification systems the category 'unclassified' refers not only to non-residential neighbourhoods but also neighbourhoods comprising institutions such as prisons, boarding schools or monasteries that do not contain normal household structures.

Figure 4.7 is a choropleth map where the style of shading indicates to which geodemographic cluster each of the 834 census areas belongs. Moreover, the density (darkness) of shading indicates the relative amount of a certain attribute within each neighbourhood group. The attribute is the proportion of respondents in each geodemographic cluster who indicated, during a hypothetical survey, that they read the *West Country World* newspaper. The darker the shading, the greater the proportion of the local population that reads the newspaper.

The proportions were determined in five stages. First, a boundary file of Bristol was obtained that gave a georeference (an ID) for each of the 834 localities. That file was loaded into the GIS and mapped on-screen – a visual check for any obvious errors it might have contained. As no errors were apparent, the second stage was to join the boundary file to a table of values containing both the same set of georeferences (the same IDs) and also a geodemographic attribute code for each area, as obtained from the Leeds classification system.

The third stage involved grouping together the 834 census areas into their nine (plus 'other') neighbourhood cluster groups (recalling that there is no rural type to be found). This process of aggregation, also known as dissolving, is illustrated by Figures 4.8(a) and (b). In Figure 4.8(a) there are 18 neighbourhood features each with a goedemographic attribute value indicating their neighbourhood classification (this attribute is represented by the shading). The result of aggregating by neighbourhood type is illustrated by Figure 4.8(b). Although the geography of the neighbourhoods appears unchanged, there are now only six features, labelled I to VI – one per neighbourhood type. The fact that map feature I is disjointed is irrelevant; it is now wholly one object. To use a physical analogy, where previously there were 18 islands, there are now six archipelagos.

The fourth stage was to overlay the aggregated neighbourhood objects upon the survey data at the same time requesting the GIS to calculate how many respondents live within each neighbourhood type (an

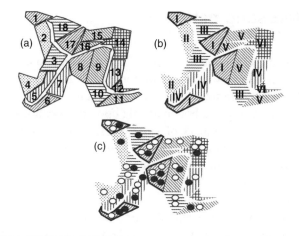

Total Surveyed = white counters + black counters = 43
Total black counters = 17
Proportion of counters that are black = 17/43 = 39.5%

Total Surveyed in Type I neighbourhoods = 15
Total black counters in Type I neighbourhoods = 7
Proportion of counters black in Type I neighbourhoods = 7/15 = 46.7%

Type I has 46.7/39.5 = 1.18 more black counters than average
Index value for black counters in Type I neighbourhoods = 1.18 × 100 = 118
(An index value of 100 is average).

Figure 4.8 Calculating the proportion of respondents per neighbourhood type. (a) and (b) Aggregating the neighbourhoods together into geodemographic clusters. (c) Calculating the proportion of respondents per cluster. In this example, the black counters indicate a survey respondent

operation known as a point-in-polygon analysis) and, of those, how many read the *West Country World*. This stage is illustrated by Figure 4.8(c).

To complete the analysis and to map the survey data requires the locations of respondents be given a geographical coordinate consistent with the referencing system used in the boundary file. In the Bristol example the coordinate system is the British National Grid. Either of UK Ordnance Survey's Address-Point or Code-Point data products could be used to assign a National Grid reference to the residential (or business) address of survey respondents (see Chapter 3, Section 3.1 for details). Having calculated the number of sample respondents within each neighbourhood type it is a simple undertaking to calculate the proportion that read the newspaper per neighbourhood group and consider whether the proportion is above or below average for the region (Figure 4.8(c)).

4.3 An interesting pattern?

Let us assume that the hypothetical survey had managed to contact 2835 households – approximately 1.5% of all households in Bristol. Of these, imagine 577 (20.4%) were found to read *West Country World*. The GIS is used to overlay the aggregated boundary file upon the locations of survey respondents finding the numbers of survey respondents in each neighbourhood class. The results are shown in Table 4.3 and suggest that the newspaper is most popular in Type 3 neighbourhoods and least popular in Type 5 neighbourhoods.

Suppose that Type 3 neighbourhoods are the 'Settled Suburbans' group of the 1991 SuperProfiles classification (see Figure 1.5). This group is described by Brown and Batey (1994, Table 3) as follows.

These families are well established in their semi-detached suburban homes. The Settled Suburbans are employed in white collar and middle management positions. The presence of many part-time working wives ensures a fairly affluent lifestyle. For example, this group can afford to take one or two packaged holidays every year and purchase newer cars. They have taken advantage of government share offers in the past and are happy to use credit cards for their purchase. Many are mail order agents. Typical publications read include *The Daily Mail*, *The Express*, *Ideal Home* and *Family Circle*.

The description is known as a pen picture (or portrait) – an evocative verbal description that highlights the special features of each neighbourhood cluster that differentiates it from the rest (see Chapter 6, Section 6.8 for a discussion of some of the advantages and disadvantages of this type of labelling). In the example case, the geodemographic analysis of the survey data can help the editors of the newspaper understand their core readership and position the content accordingly.

Or, can it? The answer, in fact, is no! The survey data segmented by neighbourhood type in Table 4.3 were actually a set of randomly generated locations. This is not to suggest there are no patterns present in the data; just that they are insignificant. Any apparent geodemographic patterns are entirely due to a random process of selection. This raises an important caution to all sorts of geographic analysis: there is a risk of attributing importance to apparent spatial patterns in geographical information when the patterns actually are of little significance. A chi-squared (χ^2) statistical analysis of the observed relative to the expected counts in Table 4.3 suggests that there is a 46% likelihood that the geodemographic distribution of the newspaper's sample readership is due to chance. By contrast, the distribution shown in Table 4.4 has less than 1% likelihood.

Table 4.3 Geodemographic analysis of the survey data. Neighbourhood patterns in the data are actually due to a random process of selection

Cluster group (k)	Total number of survey respondents (n_j)	Read West Country World (observed, O)	Read West Cty. World (p, %)	Ratio $p : \bar{p}$	Relative index value = $p/\bar{p} \times 100$	Rank	Read West Country World (expected, E) = $n_j/\Sigma n \times \Sigma O$	$(O - E)^2/E$
1	394	82	20.8	1.02	102	4	80	0.03
2	887	191	21.5	1.06	106	3	181	0.55
3	609	138	22.7	1.11	111	1	124	1.51
4	127	25	19.7	0.96	96	5	26	0.03
5	108	16	14.8	0.73	73	9	22	1.66
6	57	10	17.5	0.86	86	6	12	0.23
7	51	11	21.6	1.06	106	2	10	0.03
8	295	49	16.6	0.81	81	8	60	2.09
9	0	0	0	–	–	–	0	–
10	211	37	17.5	0.86	86	7	43	0.85
	$\Sigma n = 2739$	$\Sigma O = 559$	$\bar{p} = 20.4$	1.00	100		$\Sigma E = 559$	$\chi^2 = 6.98$
Other	96	18						$(p_{chance} = 46\%)$
	$\Sigma = 2835$	$\Sigma = 577$						

Table 4.4 Geodemographic analysis of alternative survey data. Neighbourhood patterns in the data appear much less likely to be due to chance

Cluster group (k)	Total number of survey respondents (n_j)	Read West Country World (observed, O)	Read West Country World (p, %)	Ratio $p : \bar{p}$	Relative index value = $p/\bar{p} \times 100$	Rank	Read West Country World (expected, E) = $n_j/\Sigma n \times \Sigma O$	$(O − E)^2/E$
1	394	87	22.1	1.08	108	4	80	0.54
2	887	198	22.2	1.09	109	3	181	1.59
3	609	150	24.6	1.21	121	1	124	5.32
4	127	21	16.5	0.81	81	5	26	0.93
5	108	10	9.3	0.45	45	9	22	6.58
6	57	9	15.8	0.77	77	6	12	0.60
7	51	12	23.5	1.15	115	2	10	0.24
8	295	41	13.9	0.68	68	8	60	6.13
9	0	0	0	–	–	–	0	–
10	211	31	14.7	0.72	72	7	43	3.38
	$\Sigma n = 2739$	$\Sigma O = 559$	$\bar{p} = 20.4$	1.00	100		$\Sigma E = 559$	$\chi^2 = 25.31$
								($p_{chance} \approx 0\%$)

Although superficially similar to Table 4.3 the underlying process of selection no longer was entirely random.

4.4 Confounded by choropleths!

Being led to misinterpret patterns in geographical information is a particular problem with choropleth maps. Figure 4.9(a) shows again the Canadian income data previously mapped in Figure 4.5. Figures 4.9(b) and (c) plot the *same* data but modify the way the regions of the map are grouped together and shaded (the changes are indicated by the legends). It can be argued that different handling of the same data alters the geographical knowledge and understanding of the income distributions which are read from the map. This circumstance relates to the old adage that 'a picture tells a thousand words'. The visual characteristics of a map are extremely persuasive, so much so that they can lead to a sometimes false sense of security in the objectiveness, neutrality

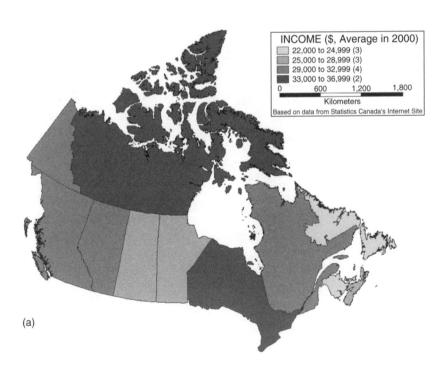

(a)

Figure 4.9 Income maps of Canada. (a) Classification by nested averages

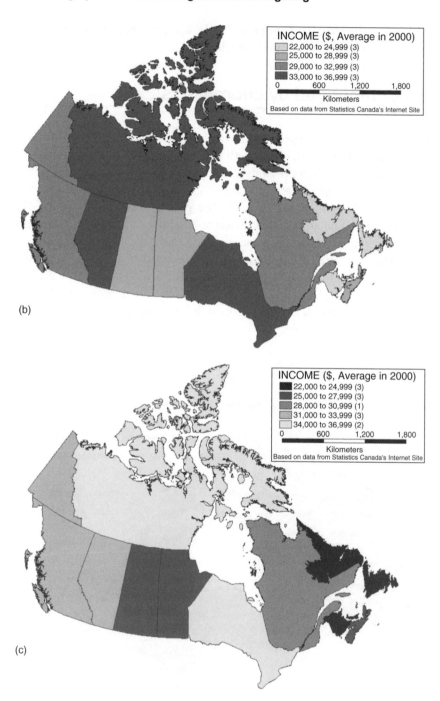

Figure 4.9 (*Continued*) (b) Classification by quartiles. (c) Classification by equal interval bin widths

and 'truthfulness' of the map and its designer. In practice maps, like statistics, can be manipulated to paint very particular pictures (Pickles, 1992; Harley, 1992; Monmonier, 1996) – a property that is useful but open to abuse.

Taking the maps in reverse order, Figure 4.9(c) has five classes, each with an equal class size interval. This means the difference between the upper and lower values – the 'bin width' or the range – is the same for each class, in this case a difference of 2999 Canadian dollars (i.e. 36,999 − 34,000, or 33,999 − 31,000, etc.). Holding the bin widths constant provides no guarantee that there will be the same number of map features (the provinces and territories) in each class. Indeed, the distribution across the classes is 3, 3, 1, 3 and 2.

By contrast, Figure 4.9(b) does have an equal number of features in each class. It is a quantile classification and, specifically, a quartile grouping (since there are four classes; a decile classification would have 10 classes and a quintile classification would have five). To accommodate this (and because the map features are not uniformly distributed across the income range) the bin widths are no longer equal. It is only possible to place an exactly equal number of features into each class if the number of features is an integer multiple of the number of classes (in other words, dividing the number of features by the number of classes gives a whole number and no remainder). If not the case, as when 13 features are grouped into four classes, the GIS will produce a near fit, for example 3, 3, 3 and 4.

Figure 4.9(a) has been produced by looking for natural breaks in the data. A number of algorithms exist to do this and in this case it is by a process of nested averages. The features are first split into two classes: above and below the mean average income for all provinces and territories. The average income is then found for each subset of features in a class and the classes again bisected at the average for the subset. This process can continue to create 2, 4, 8, 16, 32, (etc.) classes. There may be neither equal bin widths nor an equal number of features in each class.

There are many other ways we could have classified, symbolized and mapped the income data. Each would have given a different impression of the distribution of incomes across Canada. Monmonier (1996, pp. 1–2) comments 'a single map is but one of an indefinitely large number of maps that might be produced for the same situation or from the same data.' He warns that 'because of personal computers and electronic publishing, map users can now easily lie to themselves – and be unaware of it.' Marketing experts specifically are mentioned!

Looking again at Figure 4.7 – the geodemographic map of Bristol – there are three limitations to it that are true also of the income maps, Figures 4.9(a)–(c):

- First, it is implied that populations are distributed uniformly within each area and that all zones are fully occupied by residents (there are, apparently, no non-residential spaces within each area).
- Second, it is also implied that populations are socio-economically uniform within each area division – 'one size fits all' with all residents within a neighbourhood assigned the same map label.
- Third, the map implies that all change in geodemographic condition occurs only at the borders between zones.

Each of these implications may be demonstratively false. Not all zones are uniformly populated (Figure 4.10), neighbourhood populations are not uniform in their characteristics (Figure 4.11) and geodemographic change need not be at the boundaries of administrative zones (Plate 2). The limitations of the choropleth map as a tool to visualize neighbourhood data helps explain why point maps are often used instead in the sorts of geodemographic information systems that we discuss in the following chapter.

Figure 4.10 By using a GIS to overlay the boundaries of 2001 UK Census output areas onto a remotely sensed image of a coastal region it can be seen that not all the Census OAs are uniformly populated

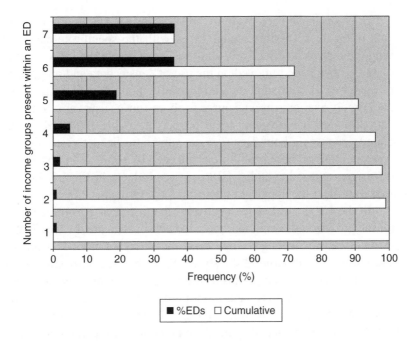

Figure 4.11 Within Bristol, over 90% of 1991 Census EDs were represented by at least five different household income groups from a total of seven in the range less than £5000 to over £40,000 per annum. Source: Adapted from Harris and Longley (2002)

Case study: Using GIS for neighbourhood analysis and targeting – a commercial perspective

Stewart Berry, GIS specialist, Caliper Corporation

There are an almost overwhelming number of geographic data and software products that cater to those wishing to differentiate target populations according to the neighbourhoods in which those populations live. Applications of these technologies are diverse and include marketing and direct mailing, and the identification of deprived communities or 'at-risk' populations.

The primary source of neighbourhood information for these products is typically a national census. This usually provides the most important government data for neighbourhood analysis due to its scope and the wealth of descriptive variables that are enumerated. Associated boundary information, produced to support and enable the collection of the census data, also facilitate subsequent analysis.

Big steps have been made in the USA and the UK to provide census data in easily accessible media, often with the specific intention of allowing the data to be effectively analysed in a GIS. In the USA, restrictions largely have been removed on the use of census data and on governmental geographic data such as administrative boundary files. The data are made freely available as a resource for both census users and commercial data providers. Census data are also freely available in the UK but the associated geographic boundary files are subject to licence restrictions if used commercially. These prohibitions are seen by some as an obstacle to market driven and value-added GIS census products in the UK but are required to protect the copyright interests of the Ordnance Survey.

In the USA and UK (respectively) the principal government portals for the provision of neighbourhood and census data are the GeoSpatial One-Stop (www.geodata.gov/gos) and the Neighbourhood Statistics website (http://neighbourhood.statistics.gov.uk; but see also Chapter 3, Section 3.2). Whether such portals also offer the necessary ease of access required by some users may be a moot point. This is because many GIS users do not want to have to wade through different websites or to order multiple CDs to get the data they need, and *then* have to spend time getting it into their preferred mapping system in a suitable format.

Consequently, and despite the progress in making census outputs available to users, obtaining, mapping and joining GIS boundary files to other neighbourhood data may still prove to be a time-consuming process (see Figure 4.12(a)). Not only can the digital boundaries be non-coincident with the user's target area of interest, the resolution of the data also varies depending on the source or version. For example, the 2001 UK Census output areas have been made available as either high-resolution or low-resolution boundaries and in such cases the user needs to determine the most appropriate product to use.

Complicating matters further, the data can be (and often are!) in any one of several different formats due to competing commercial offerings, and further non-commercial and governmental file types. The task is made easier by GIS software from vendors such as ESRI, MapInfo and Caliper Corporation that support the major formats directly, as well as having the ability to import an extensive number of formats into their own proprietary file types. There is also other format conversion software available such as the Feature Manipulation Engine from Safe Software (www.safe.com), and the Geographic Translator from Blue Marble Geographics (www.bluemarblegeo.com).

In addition, there can be issues involving projections, coordinate systems and even datums. This is often the least understood aspect

(a)   (b)

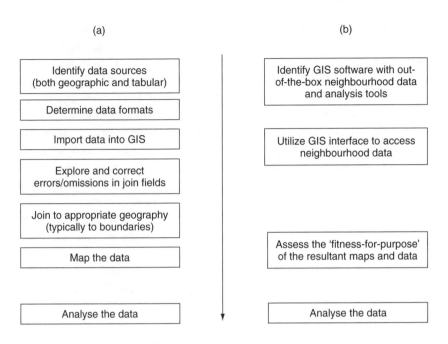

Figure 4.12 Using GIS for neighbourhood mapping and analysis. (a) A 'loose coupling' of census data with a GIS. (b) An integrated provision of census data with a GIS

of integrating data in a GIS, mainly due to the specialized knowledge that is required. A good starting point is *Datums and Map Projections* (Iliffe, 1999).

The census data themselves are not problem-free. Although a valuable source of information that is generally agreed to be the most reliable and complete population coverage available, spatial and temporal inaccuracies exist which are caused by undercounts (particularly in the USA where a mail out, mail back survey method is used) and the infrequency of collection (typically decennial but every five years in countries including Canada). Census surveys also do not always ask the questions of most interest to some users – notable in this respect is the absence of an income question in the UK. Additionally, those identifying target market populations generally are not interested in the variations described by the 'raw' census statistics but prefer to consider matters in terms of neighbourhood typologies such as 'Educated Urbanites'.

Due to such considerations, commercial databases have been developed that augment census datasets. These include population projections for non-census years and neighbourhood (geodemographic) classifications. With regards to the former and in the UK, the Joint Industry Committee on

Population Statistics (JICPOPs), a media industry trade group, has imposed a requirement of rival geodemographic suppliers to use consistent annual population estimates for postcode sectors. The latter are derived from the modelling of population distributions by applying cluster and principal components techniques to aggregate census data (cf. Chapter 6). Classifying neighbourhoods in this way obscures underlying variability in the population.

A different approach is to use consumer-derived databases that describe the attributes of individuals or their households ('lifestyle' products, see Chapter 9). This information is typically obtained from sources other than a census, including shopping surveys, share ownership data, court judgements and consumer goods warrantees (Birkin, 1995). Although using lifestyle data can in some applications sidestep the problem of the ecological fallacy – the problem of inferring the characteristics of individuals from the characteristics of places – the use of lifestyle data has its own problems, most notably that the data are partial, in the sense that they are only a sample of the complete population. The various geodemographic and lifestyle products available are discussed elsewhere in this book (see, in particular, Chapter 3).

Frustration with identifying and preparing the required census data for use in a GIS (Harris, 2003), coupled with the difficulty of 'unlocking' the perceived value of census data, has led some to look to GIS systems that include value-added census and other information in an integrated environment. The vendors of these GIS products have made steps to move away from generic GIS tools towards out-of-the-box 'total solutions', such as MapInfo's PSYTE products which come with clustering and customer identification tools, and Microsoft's MapPoint software that is supplied with demographics and consumer purchase behaviour information. Caliper Corporation facilitate access to the hundreds of tables in the US and UK censuses through their Table Chooser Tools that group the tables by category such as gender, income (US only) or age, each with associated sub-tables such as gender by age. These variables are then mapped via the tool and Caliper's GIS system (Maptitude) to the user's target area boundaries such as: tracts (US); blocks (US); output areas (UK); or postal geographies (US and UK).

Provided above is a simplified description of the steps required to map neighbourhood data in a GIS. Figure 4.12(a) shows the procedure for integrating external data with a GIS. Figure 4.12(b) describes the use of a purpose-built interface to access the neighbourhood data, simplifying the process but limiting access to the built-in data. Of course, each method could be used in conjunction with the other, subject to the task at hand, the GIS employed, the available data and the expertise of the user.

Deciding which data and system to employ and from where to obtain them are the first steps in the process of neighbourhood analysis and targeting. While more complex issues such as the modifiable areal unit problem (MAUP, Chapter 8) and the ecological fallacy (Chapter 2) should always be considered, concerns about price, usability and 'fitness-for-purpose' can be the critical factors in a project where commercial priorities may outweigh what are considered more academic concerns.

4.5 Spatial interaction models

Before concluding this chapter it is instructive to consider how the following retail question might be answered within a GI analytical environment. Suppose a consumer lives at location i (to be found at coordinate x_i, y_i) then what is the likelihood that they will visit a particular retail store at some other location j (at coordinate x_j, y_j)?

To answer this question, two ideas seem intuitive. First, that the attraction of the store to the consumer depends on what the store has to offer. This we shall quantify as the store's mass (M_j), for reasons that will become clear. Second, that the consumer would prefer to travel a shorter distance to visit a store than a longer one and so the attraction of the store is related to the distance between locations i and j. These two assumptions allow the following spatial interaction model to be formed:

$$F_{j(i)} \propto \frac{M_j}{(d_{ij})^c} \qquad (4.1)$$

where $F_{j(i)}$ is the 'force' of attraction that the store at j exerts on the consumer at i, increasing with the store's 'mass' but decreasing with the distance between i and j (the symbol $\propto$ means proportional to). The power function, c, controls the distance. The higher its value, the less attractive the store is to a consumer at a far distance relative to a consumer closer to the store. If c is given the value two and a constant g is introduced then Equation (4.1) can become:

$$F_{j(i)} = g \frac{M_j}{(d_{ij})^2} \qquad (4.2)$$

Equation (4.2) is now analogous to a simple gravity equation expressing the force acting on an object i that is due to a body of mass M_j at distance d_{ij} from i.

So far only a single customer has been considered. S/he would have to have deep pockets to ensure the long-term success of the store! Extending Equation (4.2), the total attraction of the store for n customers could be found by calculating and summing together the attraction for each individual consumer in turn:

$$\sum_{i=1}^{n} F_{j(i)} = \sum_{i=1}^{n} g \frac{M_j}{(d_{ij})^c} \tag{4.3}$$

(The symbol Σ means sum of; for example $\Sigma_1^4 = 1 + 2 + 3 + 4$)

However, recalling the geodemographic-based model of price sensitivity in a restaurant market described as a case study in Chapter 1 and also the model of survey respondents per neighbourhood type illustrated by Figure 4.8, it is expected that consumers living in a particular type of neighbourhood may have a different demand for the store compared to consumers living in another neighbourhood type. If the type of neighbourhood is denoted as k then neighbourhood differences in demand can be incorporated by giving each type a separate value of g and c. If there are n_k consumers in neighbourhoods of type k then the total attraction of the store to them is:

$$\sum_{i=1}^{n_k} F_{j(i)_k} = \sum_{i=1}^{n_k} g_k \frac{M_j}{(d_{ij})^{c_k}} \tag{4.4}$$

But, given different numbers of consumers per neighbourhood type, then a better basis to compare neighbourhoods is with regards to the average attraction:

$$\overline{F}_{jk} = \frac{\sum_{i=1}^{n_k} F_{j(i)_k}}{n_k} = \frac{\sum_{i=1}^{n_k} g_k \frac{M_j}{(d_{ij})^{c_k}}}{n_k} \tag{4.5}$$

Equation (4.5) is less complicated than it may first appear! Remember, it only formalizes the assumptions made: that consumers are more likely to travel to a store with larger 'mass'; that consumers are unlikely to travel to a store that is too far away; that demand for the store varies by neighbourhood type; and so too does the distance decay function (it is known, for example, that poorer neighbourhoods tend to do more of their shopping locally than richer neighbourhoods).

A measure of mass could be the size of the store or the number of specialist product lines it sells, while a survey could be used to measure differences in demand by neighbourhood (as in the restaurant case study).

Distance is more ambiguous. The shortest, straight-line distance between any i and j is easy to calculate given knowledge of their x and y locations (d_{MIN}: Equation (4.6)) but is an unrealistic measure if that shortest path cannot actually be traversed. Travelling from one to the opposite corner of a block in a city built to a grid plan often requires travelling east/west then north/south (or vice versa) where the diagonal path is blocked by buildings. In this circumstance the 'Manhattan distance' could be a more realistic measure of the length travelled (d_{MAN}: Equation (4.7)). Moreover, often what is meant by distance is really accessibility, in which case the 'distance' between i and j is a function of the road and other transport networks, their length and the speed of travel along them – a speed that will likely change with the time of day and with the season. It may also be that consumers travel to the store not from home but from their workplace.

$$d_{MIN} = \sqrt{((x_j - x_i)^2 + (y_j - y_i)^2)} \qquad (4.6)$$

$$d_{MAN} = |x_j - x_i| + |y_j - y_i| \qquad (4.7)$$

(The vertical 'brackets' $|\ldots|$ mean find the absolute difference between the two quantities, so ignore the negative sign if the subtraction creates a value less than zero. For example, if $x_j = 10$ and $x_i = 20$ then $|10 - 20| = 10$ *not* -10. Similarly, if $y_j = 30$ and $y_i = 50$, then $|30 - 50| = 20$. The Manhattan distance from i to j is therefore $10 + 20 = 30$, which makes more sense than saying we travelled a (negative) length of -30 units.)

Issues of how to quantify the components of Equation (4.5) are neither trivial nor insurmountable but require further assumptions (simplifications of reality) to be made. Having done so, Equation (4.5) could be used to estimate the attractiveness of a series of stores and that information used to calculate potential profitability, or to identify over- or underperforming stores. It could be used to identify potential locations for new stores on the basis of the local neighbourhood profiles provided by a geodemographic classification and it could also be used to estimate the impact on existing stores of a new store opening. For example, if a store is opened at the centre of a neighbourhood type that has above-average likelihood of residents visiting that particular chain then it is unlikely that those residents will travel further distances to visit existing stores. If the average spend of residents of the particular neighbourhood type is known and an estimate of how many of the local population travel to existing stores can be provided, then the impact of the new store on the existing network can be modelled.

Whether the simplifying assumptions of the model are realistic or not depends a great deal on the context. There may be external influence (externalities) that lead consumers to prefer one shopping centre over another. Consumers may not act as rationally as the model assumes. The impact of competitor stores and their marketing strategies will almost certainly be relevant. All these additional factors suggest that our simple interaction model has potential for considerable development. For a chronological review Birkin, Clarke and Clarke (2002) suggest: Wilson (1974); Fotheringham and O'Kelly (1989); Sen and Smith (1995); and Fischer and Getis (1999).

Some words of caution apply. In 1996, Birkin *et al.* (pp. 71, 72) wrote 'there is no effective way in most GIS of estimating the new store revenue in light of the competition' and 'proprietary GIS are often too inflexible to handle the variety of real decision-making environments that the complex modern retail environment demands.' In the following decade most of the same authors (Birkin, Clarke and Clarke, 2002, p. 151) were still of the view that

> there are few commercial packages available to enable a retail analyst to run a suite of spatial interaction models [. . .] The analyst could use one of the proprietary models available in some GIS packages [. . .] However, these models tend to be very aggregate (and simple) models and prove to be poor predictors in complicated retail markets.

4.6 Conclusion

As the name implies, geodemographics is the coupling of information recording location (the geo) with other demographic, attribute data. This coupling defines a geographical dataset and, since GIS are systems to capture, store, transform, analyse, and display geographical data, it should not be surprising that GIS and geodemographics can be stablemates.

Geodemographic mapping is aligned to vector GIS which have their origins in cartography. Under the vector model, neighbourhoods are usually represented as two-dimensional areas – as polygons. Mapping geodemographic data usually requires a boundary file to be obtained to represent the neighbourhood and to which the attribute information is joined. Other GIS functions such as an overlay, aggregation or a point-in-polygon analysis may be required. The result may reveal apparent geodemographic patterns in a sample dataset. However, there is a risk that the patterns are due to chance and so the significance of any result should be tested. Maps can lie!

The process of collecting, managing and analysing neighbourhood information can be time consuming and difficult. Recognizing this situation and the market opportunity it presents, a number of GI software and data vendors have begun offering more joined-up solutions, making the process of acquiring, handling and visualizing neighbourhood data easier. Such systems can be extremely useful for all sorts of geographical enquiry and the cost of purchase is reducing. For example, at the time of writing a single-user licence of Maptitude GIS was priced at $495 (£400 for a UK version).

GIS have their critics (limited spatial analysis, 3D visualization and temporal modelling are some of the complaints) but are nevertheless widely used to solve real-world problems and 'to improve many of our day-to-day working and living arrangements' (Longley *et al.*, 2001, pp. 30–1). Longley *et al.* (1999) give examples of GIS applications in: the utilities; telecommunications; transportation management; emergency management; land administration; urban planning; the military; public libraries; health care; politics; monitoring land cover and land use; landscape conservation; agriculture; environmental assessment and even rebuilding a country!

Yet this 'galley of applications' also hints at why GIS could be a 'nearly' technology for marketers. GIS are powerful and flexible, perhaps too much so, offering more functionality but, consequently, requiring more expert knowledge of how to use the software than a 'typical' geodemographic user may actually require. In this light and again spotting a market niche, geodemographic vendors have developed their own analysis and visualization systems – what we, in the following chapter, call a geodemographic information system (GDIS).

Summary

- GIS are software used to capture, store, transform, analyse and display geographical data.
- A geographic dataset contains both attribute and location information. The act of georeferencing adds geography to data.
- A common georeference permits tables of data to be related and jointly manipulated.
- The vector model commonly represents real-world objects as points, lines or polygons that are defined by a single or series of point coordinates.
- A raster can be understood as a grid with associated attribute values.
- GIS functions connected with geodemographic types of investigation include aggregation, overlay and point-in-polygon analysis.

- There is a risk of attributing importance to apparent spatial patterns in geographical information when the patterns actually are of little or no significance. A related problem is that maps can 'lie'.
- Despite their many proven applications, GIS have been described as a 'nearly technology' in marketing.

Further Reading

- Birkin, M., Clarke, G. and Clarke, M. (2002) *Retail Geography and Intelligent Network Planning*, Wiley, Chichester.
- Longley, P., Goodchild, M., Maguire, D. and Rhind, D. (2001 and 2005) *Geographic Information Systems and Science*, Wiley, Chichester. Supplementary material at www.wiley.co.uk/gis.
- O'Sullivan, D. and Unwin, D. (2003) *Geographic Information Analysis*, Wiley, New York.

5
Geodemographic Information Systems and Analysis

Learning Objectives

In this chapter we will:

- Introduce what we mean by the term geodemographic information system (GDIS).

- Compare and contrast our characterization of GDIS with conventional understandings of geographic information systems (GIS).

- Discuss the types of data input, analysis and output functions associated with a GDIS.

- Provide worked examples of neighbourhood profiling and catchment analysis undertaken with GDIS software.

- Give a case study of how geodemographic profiling has been used to align styles of policing with types of neighbourhood.

- Highlight the risk of making inappropriate inferences about populations when using a neighbourhood-based approach.

Geodemographics, GIS and Neighbourhood Targeting   Richard Harris, Peter Sleight and Richard Webber
© 2005 John Wiley & Sons, Ltd   ISBNs: 0-470-86413-3 (HB); 0-470-86414-1 (PB)

Introduction

In the preceding chapter we took the systems view of geographic information systems (GIS), describing such technologies as a set of linked components used to capture, store, transform, analyse and display geographical data (after Haggett, 2001, p. 719). In early GIS each of these components might exist as separate programs, with the output from one forming the input to another. Today users expect much greater integration and interoperability, and desktop GIS have become powerful and widely used tools for managing a wide range of geographical information.

However, that entire range is not required for geodemographic types of analysis such as neighbourhood classification and targeting. With geodemographics the focus is on collecting data about, classifying and visualizing socio-economic, demographic and consumer patterns, usually for predefined, small area geographies. Typically census or postal zones are used and in Chapter 4 we described how such interpretations of neighbourhood formally can be encoded in a GIS, treating each neighbourhood zone as a distinct and clearly bounded geographical entity. This entails an object or mosaic view of the socio-economic landscape and tends to favour the vector data model.

In Chapter 4 we also outlined some GIS functions used for analysing and manipulating neighbourhood objects within an applied, geodemographic context. These included the ability to:

- join generic GIS databases about neighbourhoods and their populations to local sources of information such as survey data or customer records;
- group zones together based on a common geography (e.g. sum together population counts for all unit postcodes in sector BS8 1);
- profile groups of zones by 'neighbourhood type';
- calculate and model a catchment area around a given point or zone and determine that catchment's neighbourhood composition;
- summarize and visualize information or analysis by means of tables and maps.

We expect these functions to be present too in what we here describe as a geodemographic information system (GDIS, after Goss, 1995). Yet, we also anticipate differences between 'regular GIS' and systems that are specifically built for neighbourhood profiling and micromarketing. This is because the aims, objectives and users of the two types of system are not the same. Whereas GIS are used in many areas of commerce, service delivery, environmental modelling and research, GDIS are understood as particularly designed for:

- identifying which customers buy what and where;
- identifying new customers based on the residential characteristics of existing ones; and
- branch and network planning, including measuring branch performance.

Undoubtedly GDIS overlap with GIS, both drawing upon the interdisciplinary pool of ideas and thinking that characterize geographical information science. And, without question, GIS can be well suited to querying, layering and mapping neighbourhood objects, as well as relating those objects to sources of neighbourhood data. Nevertheless, it is in our opinion wrong to view GDIS as simply 'slimline' GIS. To do so is to overlook the important differences in purpose and users of the two types of system. The requirements, ways of working and analytical methods used by a retail company's marketing department are not the same as those of an environmental scientist, for example, and so the rubric of GDIS will be seen to differ from that of GIS.

In this chapter, then, we contrast GDIS with more general notions of GIS, noting their commonalities but also the differences. The particular GDIS shown in the screenshots and used for analysis is Experian's Micromarketer (www.experian.com). However, our intention is *not* to provide a comprehensive review of this particular software and its functions but instead to offer a more generic description of a GDIS and its functionality, alongside worked examples of geodemographic types of analysis. Other products that we might characterize as GDIS include CACI's InSite (www.insite.info), Claritas' COMPASS (www.claritas.com) and EuroDirect's MICROVISION (www.microvision.info).

5.1 Data collection and input

GIS often contain a wide variety of geographic data types originating from many sources. Longley *et al.* (2001) suggest that data capture accounts for up to 85% of the total cost of implementing a GIS. Methods of data capture which they cite are: remote sensing (satellite and aerial observation); ground surveying; global positioning service (GPS); raster scans of maps and documents; manual or automated vector digitization; photogrammetry; and obtaining data via specialist geographic data warehouses such as the US

National Spatial Data Infrastructure Clearinghouse, http://130.11.52.184, or the Geography Network, www.geographynetwork.com.

To the list of geographic data sources, Burrough and McDonnell (1998) add carrying out a survey and interpolation (estimating unknown information at non-sampled locations from known information at sampled locations – an important component of geodemographic analyses where neighbourhood type is used as the inferential mechanism). Burrough and McDonnell (1998, p. 76) warn that

> in all cases the data must be geometrically registered to a generally accepted and properly defined coordinate system and coded so that they can be stored in the internal database structure of the GIS being used. The desired result should be a current, complete database which can support subsequent data analysis and modelling.

Consequently they view creating a GIS database as 'a complex operation which may involve data capture, verification, and structuring processes' (op. cit.).

5.1.1 Data collection for GDIS

We envisage that the typical GDIS user will want shielding from as much of the complex operation identified by Burrough and McDonnell (above) as possible! What is required is a system where the sorts of geographical information the user most needs are already encoded within the database. Such data could include some or all of the following: neighbourhood classifications; local market and consumer survey data; local market demographics; and trend indicators such as population projections.

The task of data collection could therefore be said mostly to fall upon the system vendor. However, it does not absolve the user of all responsibility because ideally the data supplied with the system would be the best match to the user's requirements. Such a match is best achieved by a two-way dialogue between the user and vendor, and is most achievable if the user (department or corporation) has clearly defined analytical goals and strategic marketing objectives, which translate well into data and system needs.

Even with such a dialogue, it is unlikely that the user will escape all aspects of data preparation and collation, because the users will likely be using the GDIS to analyse some of their own data – for example, client lists, customer data or the locations of stores. That data may not be held in one single database and may require a non-trivial process of finding, organizing, collating, encoding and cleaning data to some agreed, organizational standard (which may itself need to be determined).

During the process of organizing the data, the user is likely to encounter 'holes' (absent records), inconsistencies in the ways information has been collected and encoded, data entry error, duplication and so forth. The user may also wish to check how up to date certain information is by validating names and addresses against some suitable gazetteer (e.g. the latest electoral register but see recent problems with this approach in the UK: Chapter 9, Section 9.4.2). If the database is to be used for direct marketing, then good practice would advise flagging records of households that have requested not to receive 'junk mail' or telephone marketing. Then, having cleaned the database, the user may want to supplement it with additional attribute information purchased from data warehouses such as CACI's DataDepot (www.datadepot.co.uk) or EuroDirect's Data Exchange (www.eurodirect.co.uk). At all times the development and storage of the database must take place within the remits of data protection and privacy legislation such as European Union Directive 95/46/EC or The Personal Information Protection and Electronic Documents Act of Canada (www.privcom.gc.ca). Publicly funded institutions may also need to consider Freedom of Information Acts such as those of the USA or the UK (www.informationcommissioner.gov.uk).

With the above in mind and recalling that GDIS deal with only a subset of the information usually handled by GIS, it is easy to appreciate Longley *et al.*'s (2001, p. 206) view that 'data collection still remains a time consuming, tedious and expensive process.'

5.1.2 Input geographies

It is unlikely that the data collection tedium can ever fully be avoided by users of GDIS – and good fortune if it is! However, the more specified nature of geodemographics does make life a little simpler. Whereas GIS are designed to accommodate and convert between data recorded using multiple datums, projections and georeferencing systems, users of GDIS benefit from systems that are usually tailored to specific nation states and while most people know their residential (or business) zip-/postcodes, very few could express their location in terms of a grid coordinate system or by longitude and latitude. In other words, in most countries for which the systems are customized there exists a *de facto* standard for referencing small area units. These units can be used as the common 'brick' or building block to link data tables and to carry out analysis.

In the UK it makes sense to use the unit postcode as that brick-level geography. However, problems in doing so emerge when Longley *et al.*'s (2001, p. 80) characterization of a useful georeference is considered: it should be unique to one location; its meaning may be shared and understood; and it should persist (stay constant) through time. Unfortunately, postcodes do not stay constant: new properties are built that require new postcodes; industrial areas are redeveloped into residential properties that require a subdivision of postcodes; and, periodically, as the internal composition of settlements markedly changes so the postal geography may become antiquated, requiring more structured and wholesale change. Together, all these changes mean that though unit postcodes are probably the georeferencing system of socio-economic information most widely used in the UK, they meet only the unique and comprehensibility criteria for a georeference, not constancy.

Given a brick-level geography that is not constant over time, it is important that a GDIS checks for outdated codes and brings them up to date. Figure 5.1 gives an example of this. It shows a short Excel file recording the address and postcode of some of the institutions where one

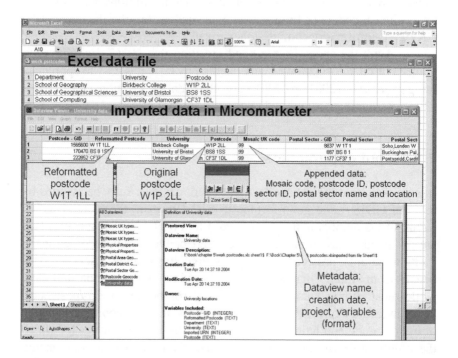

Figure 5.1 A spreadsheet with georeferencing updated and neighbourhood data appended by a GDIS

of us has worked. Below the rows of data the same information is shown, having been brought into Micromarketer. The GDIS – which incorporates postcode update information published by Royal Mail – correctly has identified that postcode W1T 1LL supersedes W1P 2LL, recording the change in the data view.

The additional and user-requested attribute data shown appended to each row include the neighbourhood type of each postcode (although, as it happens, each unit postcode is non-residential and recorded 'not classified'). Further appended is the postal sector location – information which would permit a more extensive set of data to be aggregated into these larger geographical units and joined to survey data either not available or not robust enough at the finer postcode resolutions. The software has also created metadata ('data about data') to summarize the new data view: when it was created; from which source file; and including which data.

5.2 Data analysis

A criticism sometimes made of conventional GIS is that they take a narrow view of (geographic) space, seeing it as absolute and as a container. Under this view a neighbourhood is solely defined by the innermost surface of the polygonal object that bounds it. That the place is surrounded by other neighbourhood objects is irrelevant to its own intrinsic definition because objects are deemed to exist independently of any other (Curry, 1998; see also Raper, 2000).

The perspective, which is Aristotelian and Newtonian in conception, is said to engender a false sense of separation between the users of a GIS and the world contained and represented on their computer screen. The separation in turn plays to a belief that the user is independent of and external to the apparently naturally forming objects that appear on screen and able, therefore, to gain an impartial and objective understanding of the objects, primarily by collecting data about them (Goss, 1995).

Although these and related criticisms have not always been welcomed by the GIS community (see Openshaw, 1997 for an impassioned rebuttal), they do apply in part to data analysis conducted within a GDIS. It is true, for example, that the vector-based method of defining and encoding populated zones can reduce 'neighbourhoods' merely to containers (of population). This particularly is problematic where zones are drawn around populations for administrative purposes and without consideration to the boundaries of any socio-economic and demographic phenomena the zones ultimately are used to measure. This issue is discussed further in Chapter 8.

However, it is important to recognize that the absolute view is not the only one to permeate geographic data-handling systems. In fact, analysis tends to be relational: vector data structures usually also store the topological (spatial) relationships between objects because this speeds up geographical enquiries of the database; tables of data are related and joined, combining datasets to create new information; zones are associated with larger zones higher up a census hierarchy; mapping crime rates highlights areas with the highest incidences, relative to the rest; and so forth.

The relational view especially is important in geodemographics: neighbourhood clusters are formed by grouping areas that appear most alike; a ranking by deprivation score of census wards says that higher ranked wards are areas of greater deprivation, relative to the lower; a catchment area only makes sense in terms of what it is the catchment for, be it a school, store or whatever. These and the examples that follow highlight the importance of expressing the relationships between vector-encoded point, line and polygonal objects in geodemographic types of analysis. Whereas conventional GIS offer a variety of geometric, spatial, geostatistical and 3D analysis, within a GDIS the emphasis is on assigning data to and comparing neighbourhood profiles.

5.2.1 Profiling and comparing datasets by neighbourhood

Table 5.1 provides illustration of why relationships are important and how the best matches are the most enduring! It shows by Mosaic group the number of households responding to a third-party, behavioural survey investigating the types of holidays taken by residents of a large settlement located within the West Midlands region of England. The table shows that 11,651 households responded, of which 1398 (12.00%) reside in 'Symbols of Success' postcodes, 1166 (10.01%) in 'Happy Families' areas, and so forth. The 11 UK Mosaic groups have been chosen for brevity; the more extensive list of 61 UK Mosaic (neighbourhood) types could instead have been used (see Chapter 6, Section 6.7 for further details of the classification hierarchy).

For Table 5.1(a) a catchment area has been defined as all postcode sectors that contain a survey respondent. In principle these sectors could be identified by a point in polygon analysis of the households' (x, y) coordinate locations, if known. In practice there is no need. Recall,

Table 5.1 Neighbourhood characteristics of respondents to a consumer survey undertaken within the West Midlands of England. (a) The profile is relative to a catchment area defined as the postcode sectors within which respondents live. (b) The profile is relative to the distribution of households by neighbourhood group within the West Midlands. (c) The profile is relative to persons aged 15 over who live in the Scottish National Health Service (NHS) region – and is meaningless!

(a)	Survey count	Target %	Catchment (households)	Base %	Penetration	Index
A Symbols of Success	1398	12.00	33648	10.52	0.0415	114
B Happy Families	1166	10.01	31786	9.94	0.0367	101
C Suburban Comfort	2194	18.83	54307	16.98	0.0404	111
D Ties of Community	2613	22.43	75642	23.65	0.0345	95
E Urban Intelligence	370	3.18	16997	5.31	0.0218	60
F Welfare Borderline	436	3.74	16257	5.08	0.0268	74
G Municipal Dependency	592	5.08	18083	5.65	0.0327	90
H Blue Collar Enterprise	1292	11.09	30748	9.61	0.0420	115
I Twilight Subsistence	321	2.76	10024	3.13	0.0320	88
J Grey Perspectives	742	6.37	20050	6.27	0.0370	102
K Rural Isolation	527	4.52	12317	3.85	0.0428	117
Totals	11651	100.00	319859	100.00	0.0364	100

(b)	Survey count	Target %	West Midlands (households)	Base %	Penetration	Index
A Symbols of Success	1398	12.00	168168	7.66	0.0083	157
B Happy Families	1166	10.01	203734	9.29	0.0057	108
C Suburban Comfort	2194	18.83	351686	16.03	0.0062	117
D Ties of Community	2613	22.43	466186	21.25	0.0056	106
E Urban Intelligence	370	3.18	65415	2.98	0.0057	107
F Welfare Borderline	436	3.74	120065	5.47	0.0036	68
G Municipal Dependency	592	5.08	204806	9.33	0.0029	54
H Blue Collar Enterprise	1292	11.09	303186	13.82	0.0043	80
I Twilight Subsistence	321	2.76	77186	3.52	0.0042	78
J Grey Perspectives	742	6.37	122334	5.58	0.0061	114
K Rural Isolation	527	4.52	111447	5.08	0.0047	89
Totals	11651	100.00	2194213	100.00	0.0053	100

Table 5.1 (Continued)

(c)	Survey count	Target %	Scottish NHS (persons 15+)	Base %	Penetration	Index
A Symbols of Success	1398	12.00	373893	8.92	0.0037	135
B Happy Families	1166	10.01	405650	9.67	0.0029	103
C Suburban Comfort	2194	18.83	313592	7.48	0.0070	252
D Ties of Community	2613	22.43	301788	7.20	0.0087	312
E Urban Intelligence	370	3.18	235609	5.62	0.0016	57
F Welfare Borderline	436	3.74	699377	16.68	0.0006	22
G Municipal Dependency	592	5.08	285143	6.80	0.0021	75
H Blue Collar Enterprise	1292	11.09	745702	17.78	0.0017	62
I Twilight Subsistence	321	2.76	274367	6.54	0.0012	42
J Grey Perspectives	742	6.37	258899	6.17	0.0029	103
K Rural Isolation	527	4.52	299781	7.15	0.0018	63
Totals	11651	100.00	4193801	100.00	0.0028	100

that unit postcodes are the preferred brick-level geography, which is well suited to the survey data that include a postcode identifier. Given the hierarchical nature of the UK postal geography it is easy to determine, for example, that sector DY1 1 contains unit postcode DY1 1HP. Table 5.1(a) records that the catchment contains 319,859 households, of which 33,648 are in Symbols of Success postcodes, 31,786 in Happy Families, and so forth. The distributions by neighbourhood group of the survey responding and all catchment households were calculated automatically (at the user's request) by the GDIS, using data supplied with and preloaded into the system.

Relating the sum of survey households (the numerator) to the total number of households within the catchment (the base or denominator) reveals that a proportion of $11,651 \div 319,859 = 0.0364$ of the total population were surveyed (3.64%). This 'penetration rate' (P) can also be calculated by neighbourhood group. Where n denotes the survey count per neighbourhood group and N is the corresponding catchment count then:

$$P = \frac{n}{N} \qquad (5.1)$$

For the region defined by the catchment, the highest penetration is within 'Rural Isolation' postcodes (0.0428). The lowest is among the

'Urban Intelligence' (0.0218). These differences are reflected in the final column of index values (again, in Table 5.1(a)). The index (I) for any neighbourhood group is the proportion of the survey count allocated to the group (the target proportion), relative to the proportion of the catchment count allocated to same group (the base proportion), with the case of both proportions being equal given a value of 100. Where n and N are as for Equation (5.1) and K indicates the number of neighbourhood groups, then:

$$I = \frac{n/\sum_1^K n}{N/\sum_1^K N} \times 100 \qquad (5.2)$$

which is equivalent to:

$$I = \frac{T\%/100}{B\%/100} \times 100$$
$$= \frac{T\%}{B\%} \times 100 \qquad (5.3)$$

where $T\%$ and $B\%$ are respectively the target and base percentages for the neighbourhood group. Therefore the index value for the Rural Isolation group is $4.52 \div 3.85 \times 100 = 117$ and for Urban Intelligence it is $3.18 \div 5.31 \times 100 = 60$. A value above 100 indicates more of the survey respondents are of the neighbourhood group than would be expected when looking at the distribution of the population within the catchment (an index of 117 indicates 17% more), and a value below 100 indicates less than expected (an index of 60 indicates 40% less).

The index and penetration values are generated automatically by the GDIS. They compare the neighbourhood distribution of the survey data (the observed distribution) against some baseline distribution (the expected distribution). In the case of Table 5.1(a) the base is all households in postal sectors containing a survey respondent. It appears that relative to all households in the catchment, the survey has under-enumerated the Urban Intelligence group.

It may be that that the user is less interested in how representative the survey is of the immediate vicinity to respondents and more interested in how representative it is of a larger geographical region such as the whole of the West Midlands. Such a comparison is made in Table 5.1(b). The numerator – the survey data – remains the same but the base is now the neighbourhood distribution of all households living in postcode sectors within the government region of the West Midlands.

It now appears that the survey disproportionately *over*samples the Urban Intelligence group (by 7%). This finding does not contradict the previous one because it is obtained using a different base. Rather, it suggests that Urban Intelligence group are spatially concentrated into particular localities and therefore not evenly or randomly distributed across the West Midlands region. Given the spatial patterning of this and other neighbourhood groups it is to be expected that changing the analytical base will lead to non-uniform and non-random changes to the index and penetration values, and requires the base be carefully chosen.

Interestingly, Table 5.1(b) provides evidence of a socio-economic divide in terms of the groups represented in the survey (not base) data relative to the West Midlands. Reading through the neighbourhood group names it can be inferred that, with the possible exception of the elderly, it is the more affluent neighbourhoods that are oversampled and the least afflu- ent that are undersampled. This result may link to concerns of a 'digital divide' with respect to the sorts of people data vendors are most interested in collecting information about and the sorts of decisions that are made on the basis of the data they sell. However, this 'divide' is not always so apparent – see Chapter 9, Section 9.4.1).

For Table 5.1(b) the base was changed to the to the West Midlands government region, though we could instead have chosen local authority or Nielsen region, among others. However, it is not only the geography of the base that can be changed; it is also the comparative attribute data. The survey data is by household so it has been compared with other household distributions. Nevertheless, we could have chosen total population, total population aged 15 or above, or total population aged 18 or above, had these been more appropriate, comparative indicators.

For the survey data we are fortunate that a sensible base readily is identifiable. This may not always be the case and in such circumstances selecting the base will require careful consideration of the data available and the information sought from it. It is entirely possible to obtain appar- ently meaningful information describing a nonsensical relationship. Table 5.1(c) demonstrates this point by comparing the survey data with a base- line of persons aged 15 or over who live in the Scottish National Health Service (NHS) region. In this example both the geography and attribute of the base are mis-specified. Yet, results are still obtained: relative to the proportion of persons aged 15 or over within the Scottish NHS region, there is a much higher proportion of survey-responding households within the West Midland settlement that live in 'Ties of Community' postcodes. Quite what this result usefully tells us is somewhat of a mystery!

5.2.2 Profiling classed attributes by neighbourhood

An extension to the profiling methodology described above is to investigate whether different types of holiday are associated with different types of neighbourhood. The survey data have been coded with five classes of holidaymaker: those who only holiday in the UK; those who only holiday in Europe; those who holiday in the UK or Europe but not elsewhere; those who holiday in the UK and Europe, and/or outside Europe; and those who do not holiday at all. The penetration and index values can be calculated for any particular class, where the numerator is now the subset of households belonging to the class, while the base is all households in the survey. The GDIS can create multiple profiles – one for each of the classes of holidaymaker – and these will be comparable since they are each related to the same, all-survey households base.

The profiles are shown in Table 5.2 in the form of a cross-tabulation between the neighbourhood groups and the holiday destinations. Again, the cell counts and statistics are automatically generated after the user has specified the classes to compare and the cell information to include. The index values are interpreted as saying that of the households surveyed: holidaying in the UK only is mostly a characteristic of 'Welfare Borderline' neighbourhoods; holidaying in Europe only is marginally a characteristic of the 'Happy Families' group; UK and Europe destinations most characterize 'Symbols of Success' (though only narrowly more than Happy Families); holidaying outside Europe is a characteristic of Symbols of Success neighbourhoods; and no-holiday households are concentrated most in the 'Blue Collar Enterprise' and 'Municipal Dependency' areas.

5.2.3 Issues of interpretation when using the index values

Looking back at Table 5.2, careful inspection of the cross-tabulation reveals the need for caution when interpreting the index values in the ways described. Notably, the values compare only vertical differences in the table. These address the *relative* differences *between* groups but not the *actual* cases *within* groups.

Looking at the column percentages it is true that of the 1235 survey respondents where the household does not holiday, 215 (17.41%) are in Blue Collar Enterprise postcodes. This compares with 11.09% of all respondents who live in these areas, giving an index value of 17.41 ÷ 11.09 × 100 = 156.99. However, it is an ecological fallacy to assume that because *pro rata* and *en masse* the Blue Collar postcodes have higher

Table 5.2 A cross-tabulation of neighbourhood groups and holiday destinations. The shaded cells show that although no holiday appears to be a characteristic of Blue Collar Enterprise neighbourhoods, in fact the UK is a more common holiday choice within those areas

	UK only	Europe only	UK & Europe	UK, Europe &/or other	No holiday	Totals	Cell contents
Symbols	314.00	134.00	382.00	490.00	78.00	1398.00	Count
of Success	22.46	9.59	27.32	35.05	5.58	100.00	Row %
	8.29	10.53	15.48	16.97	6.32	12.00	Column %
	69.08	87.80	129.00	141.40	52.64	100.00	Cell Index
Happy	297.00	144.00	315.00	343.00	67.00	1166.00	Count
Families	25.47	12.35	27.02	29.42	5.75	100.00	Row %
	7.84	11.32	12.76	11.88	5.43	10.01	Column %
	78.34	113.12	127.54	118.68	54.21	100.00	Cell Index
Suburban	597.00	233.00	521.00	654.00	189.00	2194.00	Count
Comfort	27.21	10.62	23.75	29.81	8.61	100.00	Row %
	15.76	18.32	21.11	22.65	15.30	18.83	Column %
	83.69	97.27	112.10	120.26	81.27	100.00	Cell Index
Ties of	941.00	312.00	520.00	545.00	295.00	2613.00	Count
Community	36.01	11.94	19.90	20.86	11.29	100.00	Row %
	24.84	24.53	21.07	18.87	23.89	22.43	Column %
	110.77	109.37	93.95	84.14	106.51	100.00	Cell Index
Urban	86.00	42.00	83.00	122.00	37.00	370.00	Count
Intelligence	23.24	11.35	22.43	32.97	10.00	100.00	Row %
	2.27	3.30	3.36	4.22	3.00	3.18	Column %
	71.49	103.97	105.90	133.02	94.34	100.00	Cell Index
Welfare	210.00	51.00	59.00	51.00	65.00	436.00	Count
Borderline	48.17	11.70	13.53	11.70	14.91	100.00	Row %
	5.54	4.01	2.39	1.77	5.26	3.74	Column %
	148.14	107.14	63.88	47.19	140.64	100.00	Cell Index
Municipal	264.00	66.00	83.00	82.00	97.00	592.00	Count
Dependency	44.59	11.15	14.02	13.85	16.39	100.00	Row %
	6.97	5.19	3.36	2.84	7.85	5.08	Column %
	137.16	102.12	66.19	55.88	154.58	100.00	Cell Index
Blue Collar	552.00	138.00	188.00	199.00	215.00	1292.00	Count
Enterprise	42.72	10.68	14.55	15.40	16.64	100.00	Row %
	14.57	10.85	7.62	6.89	17.41	11.09	Column %
	131.41	97.83	68.69	62.14	156.99	100.00	Cell Index
Twilight	142.00	30.00	52.00	49.00	48.00	321.00	Count
Subsistence	44.24	9.35	16.20	15.26	14.95	100.00	Row %
	3.75	2.36	2.11	1.70	3.89	2.76	Column %
	136.06	85.60	76.47	61.58	141.07	100.00	Cell Index

Grey	233.00	74.00	149.00	200.00	86.00	742.00	Count
Perspectives	31.40	9.97	20.08	26.95	11.59	100.00	Row %
	6.15	5.82	6.04	6.93	6.96	6.37	Column %
	96.58	91.35	94.80	108.74	109.34	100.00	Cell Index
Rural	152.00	48.00	116.00	153.00	58.00	527.00	Count
Isolation	28.84	9.11	22.01	29.03	11.01	100.00	Row %
	4.01	3.77	4.70	5.30	4.70	4.52	Column %
	88.71	83.43	103.91	117.12	103.83	100.00	Cell Index
Totals	3788.00	1272.00	2468.00	2888.00	1235.00	11651.00	Count
	32.51	10.92	21.18	24.79	10.60	100.00	Row %
	100.00	100.00	100.00	100.00	100.00	100.00	Column %
	100.00	100.00	100.00	100.00	100.00	100.00	Cell Index

concentrations of non-holidaying households relative to other neighbour-hoods, so all or even a majority of the survey respondents living in those postcodes must share that characteristic. They do not! Looking at the row percentages it is found that only 16.64% of the survey respondents living in Blue Collar neighbourhoods do not have holidays – a value that is greatly exceeded by the 42.72% that holiday in the UK and nearly matched by the 15.40% that holiday in the UK, Europe and/or elsewhere.

A second limitation when interpreting the index values (but not revealed by the cross-tabulation) is the difficulty in gauging the *significance* of apparent differences between neighbourhoods (although note the *z*-score column in Table 3.5). The index values only compare the aggregate differences *between* the neighbourhood groups and they could conceal a complex pattern of holiday choices *within* the groups. In general terms, the more diversity there is within groups, the less significant the differences between groups are.

Admittedly, practical difficulties would emerge were a measure of diversity to be calculated, for example to look at differences between unit postcodes within a group. Given the survey has an overall penetration rate of 3.64% and knowing there is an average of about 15 households per post-code then this suggests an average of about one survey record per post-code. Therefore, for the average postcode the number of the population that is characterized by any particular attribute is constrained to be either zero or one (0 or 100%). Consequently, small-number effects will lead to extremely unreliable postcode-level estimates. Yet, without a measure such as the standard deviation between postcodes (see Chapter 6, Section 6.3.1 for an explanation of standard deviation), the interpretation of the index values rests on the ability of the clustering methodology accurately to define neighbourhood groups that place like population with like.

A third problem with the index values is their tendency to give disproportionate attention to the relative overabundance of a particular characteristic at the expense of less than expected values. Greater than expectation is usually indexed as above 100 and tends to 'draw the eye'. Although in itself an apparently minor matter, it does give rise to a visualization problem when indices are plotted on a histogram, because all below-average values are constrained to fall in the finite range 0 to <100, whereas all above-average values are in the range >100 to infinity. As we shall show later in this chapter, if the index axis of the histogram has an equal length interval between 0 to <100, 100 to <200, 200 to <300 and so forth, then the histogram will almost certainly appear dominated by above-average values.

The fact that an above-average value can range to infinity also means that care should be taken to ensure sample sizes are sufficiently large to warrant the interpretation of the index values. Deceptively high index values can be obtained, notably when the sample size is small and when the expected fraction of the attribute count to be found in any neighbourhood group is low. For instance, imagine a survey where only 10 from a total of 1000 respondents were found to be of a particular neighbourhood group and where a subset of 50 of the survey respondents (from all neighbourhoods) was found to display a certain consumer attribute. If 'by chance' a single member of that subset is located in the one neighbourhood then an index value of $(1/50 \div 10/1000 \times 100) = 200$ is obtained. If the subset were only 25 in size then the index value becomes 400. And if the subset were 25 but there were initially 5000 respondents then the value is 2000 – very high for what in absolute terms is a one-off occurrence and providing a second example of small-number effects.

5.2.4 Volume profiles

A second ecological fallacy in addition to that discussed in the preceding subsection would be to assume that the results of the cross-tabulations shown in Table 5.2 are necessarily representative of a larger geography such as the West Midlands as a whole. Indeed, there is good reason to suspect they are not – we have already detected an economic divide with regards to the groups that have been over- or undersampled relative to the region's neighbourhood composition (Table 5.1(b)). One way to correct for this is to give each household responding to the survey a weight calculated as 100 divided by the 'survey vs. West Midlands (households)' index value of the neighbourhood group to which the household belongs. The weights mean that households resident in neighbourhoods with an index value less than 100 in Table 5.1(b) are given *higher* weight (>1) because as a group they have been undersampled relative to the West Midlands, whereas households in neighbourhoods with an index value greater than 100 are given *lower* weight (<1) because relatively they have been oversampled.

For any of the classes of holiday destination, the weighted penetration rate per neighbourhood, relative to the West Midlands, may be calculated as the weighted count of households allocated to the group and travelling to the particular destination, divided by the total (non-weighted) count of all households in the West Midlands assigned to the group. If w_i is the weight assigned to the ith household in a group, n_k is the number of households in the kth of K groups, N_k is the regional count of households in the same group and z_i acts as a filter, taking a value of one if the household does go to the particular holiday destination and zero otherwise, then:

$$P_k = \frac{z_i \times \sum_{i=1}^{n_k} w_i}{N_k} \qquad z_i \in \{0,1\} \qquad (5.4)$$

Similarly, the index value is the weighted count of survey respondents holidaying at a particular destination and living in neighbourhood type k, divided by all survey respondents (from all neighbourhoods) who travel to the same destination and expressed relative to the fraction of all West Midland households who reside in the same neighbourhood group, k:

$$I_k = \frac{z_i \times \sum_{i=1}^{n_k} w_i / \sum_1^K \left(z_i \times \sum_{i=1}^{n_k} w_i \right)}{N_k / \sum_1^K N_k} \times 100 \qquad z_i \in \{0,1\} \qquad (5.5)$$

As previously, these calculations are not 'hand cranked' but automatically undertaken at the user's request. The volume calculations are analogous to the simpler penetration and index values shown previously (Equations (5.1) and (5.2)) with one exception: households within a group are no longer treated equally but in accordance to the weight or 'volume' data attached to them.

As it happens, in our example the weight attached to each household within a group is the same because the weights were generated at the neighbourhood, not household level. As a consequence

Equations (5.4) and (5.5) could be simplified. However, it is often the case that genuine household-level differences will be specified. For example, the analyst may have information about the amount spent by each household per holiday and want to use this to calculate the average spend on each class of holiday for each neighbourhood group. In this example, the weight (a function of household spend) really would vary by household. The type of analysis is often known as volume profiling.

Table 5.3 demonstrates the effects of the weights on the neighbourhood profiles for two of the holiday classes: Europe and/or other; and no holiday taken. They particularly act to emphasize that surveyed households living in the Welfare Borderline and Municipal Dependency postcodes of the West Midlands are more likely than expected not to holiday. Without weighting, the opposite appears true. Table 5.4 takes the analysis a stage further and considers the distribution of lone-parent households who do not holiday, relative to the West Midlands household distribution by neighbourhood groups. Non-holidaying lone parents are also found to be especially concentrated in the Municipal Dependency and Welfare Borderline postcodes. Note, however, that the number of survey respondents per subset is low.

5.2.5 Comparing neighbourhood rankings

We know, from Table 5.4, that the highest than expected concentration of non-holidaying, lone parents is within the Municipal Dependency neighbourhoods, with a volume index of 222. This group can therefore be ranked as 1. The next highest volume index is 217 for the Welfare Borderline group, which is ranked 2 . . . and so on until the Happy Families group which has the lowest volume index and is ranked 11.

Table 5.5 compares the ranked distribution by neighbourhood group of the non-holidaying, lone-parent volume indices with two other sets of ranked indices derived from the survey data: the first indicates the neighbourhood concentrations of high-income households (gross family income estimated as greater than £50,000 per annum); the second of lower income households (<£10,000 p.a.). The highest ranked neighbourhood group in terms of high household income is Symbols of Success, while the lowest ranked are the Municipal Dependency and Welfare Borderline groups. The Welfare Borderline and Municipal Dependency groups are highest ranked with regards their relative concentrations of lower income households; Happy Families neighbourhoods are lowest ranked for such households.

Table 5.3 Volume profiles, weighting survey respondents by the over- or under-enumeration of their neighbourhood groups, relative to the West Midlands. The shaded cells show the effect of weighting is to raise the proportion of non-holidaying respondents in the Welfare Borderline and Municipal Dependency groups from below- to above-average expectation

UK, Europe &/or other	Count	Target %	Weighted count (volume)	Volume %	West Midlands (households)	Base %	Penetration	Volume Penetration	Index	Volume Index
A Symbols of Success	490	16.97	312	11.55	167972	7.66	0.0029	0.0019	222	151
B Happy Families	343	11.88	318	11.75	203718	9.29	0.0017	0.0016	128	127
C Suburban Comfort	654	22.65	559	20.68	351320	16.02	0.0019	0.0016	141	129
D Ties of Community	545	18.87	514	19.02	466163	21.26	0.0012	0.0011	89	90
E Urban Intelligence	122	4.22	114	4.22	65415	2.98	0.0019	0.0017	142	141
F Welfare Borderline	51	1.77	75	2.78	120065	5.47	0.0004	0.0006	32	51
G Municipal Dependency	82	2.84	152	5.62	204806	9.34	0.0004	0.0007	30	60
H Blue Collar Enterprise	199	6.89	249	9.20	303166	13.82	0.0007	0.0008	50	67
I Twilight Subsistence	49	1.70	63	2.32	77186	3.52	0.0006	0.0008	48	66
J Grey Perspectives	200	6.93	175	6.49	122256	5.57	0.0016	0.0014	124	116
K Rural Isolation	153	5.30	172	6.36	111105	5.07	0.0014	0.0015	105	126
Totals	2888	100.00	2703	100.00	2193172	100.00	0.0013	0.0012	100	100

Table 5.3 (Continued)

No holiday	Count	Target %	Weighted count (volume)	Volume %	West Midlands (households)	Base %	Penetration	Volume Penetration	Index	Volume Index
A Symbols of Success	78	6.32	50	3.73	167972	7.66	0.0005	0.0003	82	49
B Happy Families	67	5.43	62	4.66	203718	9.29	0.0003	0.0003	58	50
C Suburban Comfort	189	15.30	162	12.13	351320	16.02	0.0005	0.0005	96	76
D Ties of Community	295	23.89	278	20.89	466163	21.26	0.0006	0.0006	112	98
E Urban Intelligence	37	3.00	35	2.60	65415	2.98	0.0006	0.0005	100	87
F Welfare Borderline	65	5.26	96	7.17	120065	5.47	0.0005	0.0008	96	131
G Municipal Dependency	97	7.85	180	13.48	204806	9.34	0.0005	0.0009	84	144
H Blue Collar Enterprise	215	17.41	269	20.17	303166	13.82	0.0007	0.0009	126	146
I Twilight Subsistence	48	3.89	62	4.62	77186	3.52	0.0006	0.0008	110	131
J Grey Perspectives	86	6.96	75	5.66	122256	5.57	0.0007	0.0006	125	102
K Rural Isolation	58	4.70	65	4.89	111105	5.07	0.0005	0.0006	93	97
Totals	1235	100.00	1332	100.00	2193172	100.00	0.0006	0.0006	100	100

Table 5.4 The distribution by neighbourhood group of lone-parent households who do not holiday, relative to the West Midlands household distribution

Lone parent, no holiday	Count	Target %	Weighted count (volume)	Volume %	West Midlands (households)	Base %	Penetration	Volume Penetration	Index	Volume Index
A Symbols of Success	8	5.67	5	3.16	167972	7.66	0.0000	0.0000	74	41
B Happy Families	6	4.26	6	3.45	203718	9.29	0.0000	0.0000	46	37
C Suburban Comfort	14	9.93	12	7.43	351320	16.02	0.0000	0.0000	62	46
D Ties of Community	35	24.82	33	20.50	466163	21.26	0.0001	0.0001	117	96
E Urban Intelligence	10	7.09	9	5.80	65415	2.98	0.0002	0.0001	238	194
F Welfare Borderline	13	9.22	19	11.87	120065	5.47	0.0001	0.0002	168	217
G Municipal Dependency	18	12.77	33	20.69	204806	9.34	0.0001	0.0002	137	222
H Blue Collar Enterprise	22	15.60	28	17.07	303166	13.82	0.0001	0.0001	113	123
I Twilight Subsistence	5	3.55	6	3.98	77186	3.52	0.0001	0.0001	101	113
J Grey Perspectives	6	4.26	5	3.27	122256	5.57	0.0000	0.0000	76	59
K Rural Isolation	4	2.84	4	2.79	111105	5.07	0.0000	0.0000	56	55
Totals	141	100.00	161	100.00	2193172	100.00	0.0001	0.0001	100	100

Table 5.5 Derivation of Spearman's rank correlation coefficient (r_S) for 'Lone parent, no holiday' vs. 'Household income > £50 k' and 'Lone parent, no holiday' vs. 'Household income < £10 k' by neighbourhood group

Neighbourhood group	Lone parent, no holiday		Household income > £50 k			Household income < £10 k		
	Volume index	Rank 1	Volume index	Rank 2	$D^2 =$ (rank 1 − rank 2)^2	Volume index	Rank 3	$D^2 =$ (rank 1 − rank 3)^2
A Symbols of Success	41	10	385	1	81	45	10	0
B Happy Families	37	11	203	3	64	36	11	0
C Suburban Comfort	46	9	82	4	25	67	8	1
D Ties of Community	96	6	35	8	4	93	6	0
E Urban Intelligence	194	3	82	4	1	74	7	16
F Welfare Borderline	217	2	18	10	64	244	1	1
G Municipal Dependency	222	1	7	11	100	213	2	1
H Blue Collar Enterprise	123	4	33	9	25	162	3	1
I Twilight Subsistence	113	5	48	7	4	155	4	1
J Grey Perspectives	59	7	67	6	1	97	5	4
K Rural Isolation	55	8	335	2	36	55	9	1
			(A) $6 \times \Sigma d^2$		2430	(A) $6 \times \Sigma d^2$		156
			(B) $n(n^2 - 1)$		1320	(B) $n(n^2 - 1)$		1320
		Vs. lone parent	$r_S = 1 - A/B$		−0.84	Vs. lone parent $r_S = 1 - A/B$		0.88

The bottom row of Table 5.5 quantifies the level of association between paired sets of ranked values. In the leftmost case the level of association between the neighbourhood rankings of non-holidaying, lone parents and high-income households is calculated; for the right-hand case it is the level of association between non-holidaying, lone parents and lower income households. The measure of association used is Spearman's rank correlation coefficient, the value of which is denoted as r_S.

Spearman's rank correlation coefficient is a 'distribution free' or non-parametric statistical test which does not make certain assumptions about the background population from which the samples are drawn (Hammond and McCullagh, 1978). It does not assume, as a number of 'classical' or parametric tests would, that the volume indices are normally distributed around the mean average, an assumption known to be inappropriate for two reasons: first, because the range of possible values above and below the mean of 100 is not equal; second, because the indices have been sorted onto a ranked (or ordinal scale) which does not preserve the interval between each ranked value (looking at the lone-parent ranking in Table 5.5, ranks 1 and 2 have a ranked difference of 1 but an index difference of $217 - 222 = -5$; ranks 2 and 3 also have a ranked difference of 1 but an index difference of $194 - 217 = -23$).

Spearman's rank correlation coefficient (r_S) is calculated as:

$$r_S = 1 - \frac{6\Sigma d^2}{n(n^2 - 1)} \tag{5.6}$$

where n is the number of ranks, equal to 11 in Table 5.5, and Σd^2 is the sum of the squared difference between the pairs of ranks for each of the n ranked observations. The coefficient, r_S can take a maximum value of $+1$ for a perfectly positive association, obtained when the two sets of ranks are identical, and a minimum of -1 for a perfectly negative association, obtained when the sets would be identical if one were to be ranked in reverse of its present order. A value of zero implies no association.

Table 5.5 shows how a coefficient value of -0.84 is derived for the neighbourhood rankings of non-holidaying, lone parents versus high-income households, and of $+0.88$, for non-holidaying, lone parents versus lower income households. These values mean that *lower* than expected concentrations of high-income households tend to be found in areas with *higher* than expected concentrations of the lone parents (the concentrations move in different directions and are negatively related); and that *higher* than expected concentrations of lower income households tend to be found in areas with *higher* than expected concentrations of the lone parents (the concentrations move in the same direction and are positively related). The significance of the results can be evaluated from the knowledge that r_S $n - 1$ has a t-distribution with $n - 1$ degrees of freedom (Rogerson, 2001). The observed values of t are therefore -0.84 $\overline{11 - 1} = 2.32$ and 0.88 $\overline{11 - 1} = 2.78$. From these and using standard statistical tables it is found that the correlation -0.84 has about 2.5% likelihood of being due to chance; the correlation $+0.88$ has about 1% likelihood.

It is important not to draw unsubstantiated conclusions from the Spearman's correlations. While it may be said that higher than expected levels of non-holidaying, lone-parent households tend to be found in the same neighbourhood groups as lower income households, it should *not* be stated that therefore lone parents have lower incomes. The correlation coefficient only summarizes a relationship at the neighbourhood scale while to say 'lone parents households are lower income households' is to express a relationship at a household scale. To presume that a relationship found at one scale holds true at another is to commit the ecological fallacy (Martin and Longley, 1995). Even if it is known that one-half (0.5) of a neighbourhood's population are lone-parent households and also that one-half of the population are lower income households, there is no guarantee that it is the same half in both cases. It is only when the average of the two proportions exceeds 0.75 that it is possible to be certain that the majority of households are indeed both lone parents and of lower incomes.

Fortunately, for our study we can return to the survey data and find the proportion of non-holidaying, lone-parent households that are also classed as lower income households. The answer is 0.65 (65%). Often, however, socio-economic data are only available having been pre-aggregated into neighbourhood units and so the cross-tabulation at the household level that we used to find the answer will not be possible to do. A frustration with census and other aggregated datasets is that while many variables are cross-tabulated, it always seems to be the ones that are not that would be of most interest to us!

There is a final problem with the Spearman's rank methodology: what happens where the ranks are tied? Usually the average is calculated so, for example, the observations with a value of 4 in the set {1, 2, 3, 4, 4, 6, 7, 7, 7, 10} are ranked fourth and fifth and given the average rank of 4.5. Similarly, the observations with a value of 7 are ranked seventh, eight and ninth and given an average of 8. Unfortunately, Equation (5.6) is not really appropriate when there are tied ranks. The correct Spearman coefficient is obtained by calculating the Pearson's correlation, using the ranks as observations (Rogerson, 2001). Alternatively, Kendall's correlation coefficient (τ), which is better for very small samples, could be calculated with a correction factor if necessary (see Hammond and McCullagh, 1978, pp. 228–33).

5.2.6 Catchment profiling

There is a final type of neighbourhood profiling that particularly is relevant to branch and network planning – catchment profiling. An example of its use was given in Chapter 1, where it helped inform the introduction of differential meal pricing within a restaurant chain. For this the neighbourhood profile of the population located within a critical threshold distance of each restaurant was calculated and the restaurant allocated to a price band according to the prevailing demographic and socio-economic characteristics of the locality. Restaurants located within or among neighbourhoods where the residents were prepared to pay more for a meal (relative to other neighbourhood groups) were assigned to the more expensive bands.

There are three principal ways of devising a catchment around a particular store or outlet, the location of that store or outlet itself being defined as either a point (x, y) coordinate or by a polygon (or in principle by a line which makes more sense for calculating the flood zone of a river, for example). The first is to define the catchment as all postal or other administrative zones containing a certain number of people who are known to visit the store. This approach is analogous to the way the survey catchment was defined for Table 5.1(a), above. Defined in this manner it is possible that the catchment will not include the zone of the store itself, which is sensible if the outlet is found in a location such as an out-of-town retail park. Specifying a minimum number of customers per zone that must be met before the zone is considered a part of the catchment is useful, for example, to exclude chance visitors who happened to pass through the area but would not usually shop at the store.

The second way is to include within the catchment all zones within a threshold distance, d, from the outlet. This is equivalent to producing a circular buffer around the outlet with the distance from it to the edge of the catchment being of length d in all directions. The main problems with this approach are threefold. First, the specified length of d is usually arbitrary. Secondly, it leads to a sudden and also arbitrary 'crossing of the line' where the characteristics of populations at the edge but within the circle are included in the catchment's profile but those just outside the circle are not. This is the same problem as with choropleth maps that imply all change in geodemographic condition occurs only at the borders between zones and arises because the catchment is represented as a discrete vector object with apparently very definite boundaries. Thirdly, profiling the catchment assumes the outlet is equally accessible to all neighbourhoods within it, regardless of any geographical variations in the transport infrastructure to be found there.

To address the accessibility issue, in terms of road access at least, a catchment can be modelled in terms of the maximum permitted drive time to the store. Roads are encoded as a network of generalized line segments within the GDIS database and each class of road is assigned a typical speed for a vehicle travelling along it. The time, t, taken to travel from any particular node to the outlet may then be calculated as:

$$t = \sum_{s=1}^{n} l_s v_s \qquad (5.7)$$

where l_s is the length of line segment s, v_s is the speed of travel along it and n is the number of segments between the node and the outlet. The time taken depends on the assumed speed achieved along a section of road which, in turn, would vary during the day (and also by season though that is less often considered). The default speeds for Micromarketer are shown in Table 5.6. These settings can be modified by the analyst.

The three types of catchment are summarized by Figure 5.2. There a fourth type of catchment is also introduced. Superficially this is the same as the circular method. However, in calculating the neighbourhood profile of the catchment, increased weight is given to zones that

Table 5.6 Default travel speeds along UK roads in Experian's Micromarketer software

	Non-peak rush hour		Peak rush hour		Weekday off-peak		Weekend & night	
	(kph)	(mph)	(kph)	(mph)	(kph)	(mph)	(kph)	(mph)
Motorway	105	65	89	55	113	70	113	70
Primary dual	77	48	56	35	92	57	97	60
Primary single	45	28	34	21	56	35	64	40
A road dual	58	36	42	26	71	44	80	50
A road single	40	25	31	19	48	30	56	35
B road dual	34	21	26	16	42	26	48	30
B road single	29	18	21	13	35	22	40	25
Unclassified dual	23	14	18	11	27	17	32	20
Unclassified single	23	14	18	11	27	17	32	20

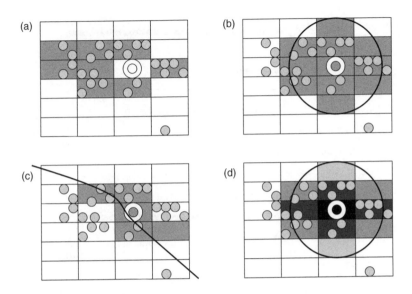

Figure 5.2 Four types of catchment area defined around a store. (a) The catchment is defined by zones containing at least two customers. (b) The catchment is defined by a circular buffer drawn around the store. (c) The catchment is defined by travel time by road to the store. (d) The catchment profile gives increased importance to zones nearest the store

are nearest to the store, and lower weight to those further away but still within the circle. This amounts to a catchment that tapers away with increasing distance from the store and therefore does not have the sharp 'is in the catchment'/'is not in the catchment' break that the first circular method has. In GIS terminology, the catchment is now modelled as a field instead of an object. This inverse distance weighting approach (as the distance from the outlet increases so the influence of the zone decreases) is often employed in types of 'hot-spot' analysis as we shall show in Chapter 9.

Figures 5.3 and 5.4 show the neighbourhood profiles of catchments drawn around hypothetical store locations within the West Midlands study region. For Figure 5.3 a three-mile catchment has been defined around two stores. In Figure 5.4 the catchment around one of those stores has been defined twice: first by road travel time at weekday peak; and second, for comparison, by weekend off-peak times. Together Figures 5.3 and 5.4 stress that the profiles of the stores depend on how their catchments are defined since the socio-economic characteristics of a store's customer base may vary at different times of the day.

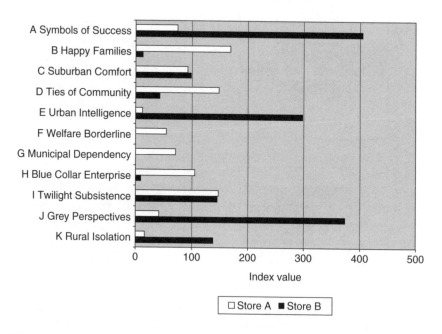

Figure 5.3 Neighbourhood profiles of the population located within three-mile catchments around two stores in the West Midlands

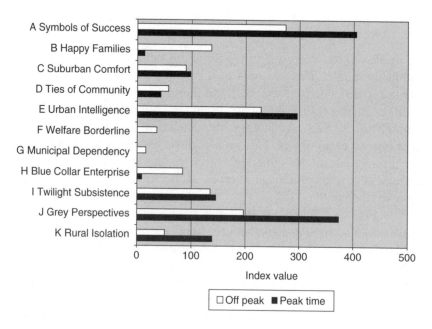

Figure 5.4 Neighbourhood profile of the population within 15 minutes of store A at weekday peak travel time and weekend off-peak time

5.3 Data visualization and output

Two of the three most common types of data summary used in a GIS have already been shown in this chapter – tabulations and graphs, usually histograms (bar charts). It has also been suggested that because below-average index values are constrained to the range 0 to 99, whereas above-average index values lie in the range 101 to infinity, then plotting index values on an equal interval histogram will lead to the below-average values being squashed in a small portion of the graph. This can be seen in Figure 5.4, where values less than 100 are plotted into one-fifth of the area of the graph, while values above 100 occupy four-fifths. Consequently the eye tends to focus upon the above-average differences between neighbourhoods.

A partial solution to this problem is to stretch the space on the graph given to values below 100 by plotting the indices on a logarithmic scale. This is shown in Figure 5.5, where the distances along the index axis are no longer of an equal interval; instead, the interval increases exponentially. Although a useful way of exploring below-average differences

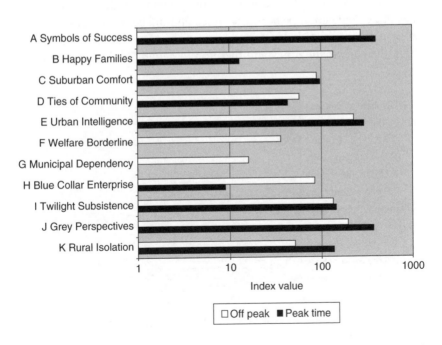

Figure 5.5 Neighbourhood profile of the population within 15 minutes of store A at weekday peak travel time and weekend off-peak time, index values plotted on a logarithmic scale

between neighbourhoods the logarithmic scale still does not give equal space to values above and below the 100 line but, in effect, reverses the previous situation by now squashing above-average values into a small part of the graph.

An alternative to Figure 5.5 is to plot the *change* in the neighbourhood profile of store A from weekday peak travel time to weekend off-peak time, as in Figure 5.6. This makes clear that the neighbourhood profile of store A's catchment is less made up of Symbols of Success neighbourhoods and more comprised by Happy Families at weekend, off-peak times *relative to* weekday peak times. Understanding that it is a relative change is important: the Symbols of Success neighbourhoods that were in the weekday catchment remain in the weekend catchment and they still have a higher than expected presence compared to the Happy Families group. What has changed is that the number of zones in the catchment rises from weekday to weekends (because traffic moves faster) and the Symbols of Success neighbourhoods constitute a lower proportion of the total.

A third type of output, possibly the one most associated with geographical information, is cartographic. Plate 3 maps the Mosaic UK groups of postcodes within a 45-minute drive time of Bluewater Shopping Centre, Kent – one of the largest shopping centres in the UK. The drive times are calculated at peak traffic times and off-peak, at weekends. The use of a 'point map' helps to retain a sense of the settlement patterns within the catchment zones and therefore a general ability to distinguish

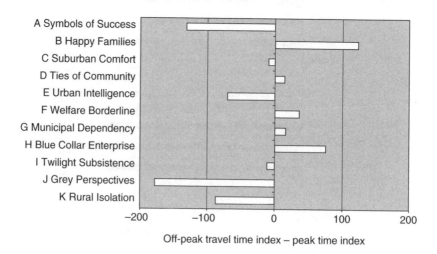

Figure 5.6 Change in the neighbourhood profile of store A from weekday peak travel time to weekend off-peak time

highly populated areas from low or non-populated ones (compare Plate 3 with the census mapping of Figure 4.7).

Visualization is an important component of presenting, exploring, revealing and explaining patterns of geographic information in a GIS (Unwin and Hearnshaw, 1994) and, in the same way, geodemographic classifications and systems often incorporate extensive visualization tools such as multimedia guides that help the user to interpret the typologies and cluster labelling. Such guides can include libraries of images, text descriptions, photographs, maps, customer profiles and other supporting material. Figure 5.7 is a collage helping to explain the characteristics of a neighbourhood type 'New Urban Colonists'. Figure 5.8 is a bar chart showing the extent to which people in different types of neighbourhood are more or less likely than average to select 'prices' as the reason why they chose the grocery shop that they do. Finally, Figure 5.9 shows what is known as a 'tree' and colours the 61 UK Mosaic types according to the proportion of survey respondents saying that they shop where they do on the basis of price. Around the border of the tree are located both demographic and attitudinal characteristics commonly associated with clusters in that area of the tree.

Figure 5.7 Screenshot from a geodemographic multimedia guide displaying a collage that helps to explain the characteristics of a neighbourhood type. Source: Reproduced by permission of Experian UK

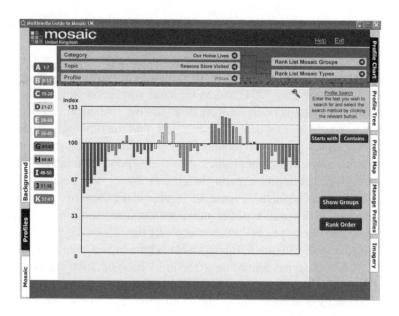

Figure 5.8 Bar chart showing the extent to which people in different types of neighbourhood are more or less likely to select 'prices' as the reason why they chose their grocery shop. Source: Reproduced by permission of Experian UK

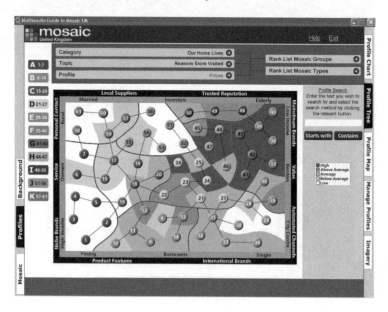

Figure 5.9 A tree visualization, colouring neighbourhood types according to the proportion of survey respondents saying that they shop where they do on the basis of price. Source: Reproduced by permission of Experian UK

Case study: Different neighbourhoods, different policing styles

Based on research undertaken by Tom Williamson, of the Institute of Criminal Justice Studies at the University of Portsmouth and by Richard Webber, in collaboration with Devon & Cornwall Constabulary. Reproduced with their permission.

The British Crime Survey currently is the leading national source of information on crime and victimization. Until recently the survey was undertaken on a biennial basis and recorded the response of some 22,000 residents to a wide range of questions ranging from: their attitudes towards the neighbourhood, the police and the criminal justice system; their experience of crime; and the response of the police to the crime.

Although the survey has been criticized for possible under-recording of repeat victimization there is no inherent reason why it should not be a reliable source of information on the relative level of victimization between different types of neighbourhood. It is for this reason that the collaborators obtained the approval of the Home Office to code year 2000 respondents according to a geodemographic classification of the postcodes in which they live – specifically, the Mosaic classification. The aim was to identify the types of neighbourhood in which respondents are most likely, for example, to have experienced theft from a car, to know the identity of an offender or to cite noisy neighbours as a serious social problem in their neighbourhood.

When the British Crime Survey results are analysed by type of neighbourhood it is clear that there is a very wide disparity between types of neighbourhood in terms of not just the overall level of crime but also the mix of crime. Neighbourhoods characterized by 'Symbols of Success' and 'Suburban Comfort' are generally likely to experience low levels of crime. Residents in these areas usually live at some distance from the types of neighbourhoods in which offenders are concentrated and benefit from vigilant neighbours and effective security. These are areas in which people tend to support the police and to help them identify offenders.

Neighbourhoods characterized by 'Ties of Community' and 'Blue Collar Enterprise' are likely to have levels of crime similar to the national average. In these areas incidents often result from drunkenness and domestic disputes. On council estates, whether low or high rise, respondents are most likely to say that they do not like their neighbourhood. These are areas where petty crime and vandalism, abandoned cars and drug addiction are cited as serious problems. Theft and burglary from dwellings are relatively more serious in these neighbourhoods than thefts from cars. Many people in these neighbourhoods know their attackers by sight but respondents say they are less likely to report crimes to the police and often do not want to get involved as witnesses.

Residents in the neighbourhood types 'South Asian Industry' and 'Settled Minorities' are the most likely to say that they suffer racial abuse and physical attacks. However, in other predominantly white areas of older terraced housing, such as 'Industrial Grit', it is evident that people feel they benefit from helpful neighbours and feel that the community is more effective in deterring criminal elements than the police.

In areas of 'Urban Intelligence', respondents complain that their neighbours are not helpful, that they (the respondents) suffer snatches of valuables in the street and that they experience break-ins to their homes. These events are perhaps due to the inner city locations of these neighbourhoods.

In areas of 'Grey Perspectives', residents like the neighbourhood where they live and look out for each other. These areas suffer low levels of crime, largely because they are distant from locations where offenders are most likely to live. Respondents to the survey who live in areas of 'Rural Isolation', if they do experience crime, are most likely to experience it in car parks or when visiting entertainment centres in cities. They feel particularly safe in their neighbourhoods, are strong supporters of the police and on average suffer very low levels of victimization. They are most likely to believe that crime is the result of poverty rather than lack of parental discipline.

As well as identifying different experiences of and attitudes to crime, neighbourhood classifications can be used to target certain types of policing within a police authority. If, for example, we want to map in Exeter the extent to which noisy neighbours is likely to be a serious problem in different communities, then the first step is to identify the neighbourhood type of each of the individual postcodes making up the town. The next stage is to look up for each postcode the score of its classification on the profile of 'noisy neighbours' appended to the British Crime Survey. If, for example, a postcode is classified as 'Town Gown Transition' and if the national index for Town Gown Transition on noisy neighbourhoods is 181, then it can be assumed that the postcode is of a sort which is more likely than average to suffer from this source of annoyance (given a value of 100 as the average, then 181 is 1.81 times greater than average). If in a local community of 20 postcodes, all 20 are of a neighbourhood type that nationally have noisy neighbours values of, for instance, over 150 (1.5 times average) then it is likely that noise will be a particular problem in the community, even if not necessarily in each and every postcode.

5.4 Conclusion

In this chapter we have provided an overview of the sorts of data input, analysis and output functions that we would expect to be present in what we have described, after Goss (1995), as a geodemographic information system (GDIS). Throughout, we have contrasted our general conception of a GDIS against wider notions of geographical information systems (GIS) and what they are used for. Recalling that GIS can be described as software used to capture, store, transform, analyse and display *geographical* data, then GDIS similarly could be described but as applying to *geodemographic* data.

Geodemographic datasets are understood as containing information about people and the places those people live in but with the data usually only available aggregated into (postal or census) zones. Accordingly, it is not actually individual people or households that are analysed and compared in GDIS but groups of populations, sorted by neighbourhood type. Often what is conducted is a method of interpolation whereby the characteristics of non-sampled persons are inferred from the fact they reside in the same type of neighbourhood as those who were sampled. Such inferences always are susceptible to the ecological fallacy but will be most successful if the underlying neighbourhood classification has successfully grouped zones of like with like and if there is clear evidence that different types of population attribute are associated with different sorts of neighbourhood.

Although geodemographic information rightly can be regarded as a subset of geographical information, it does not follow, in our opinion, that GDIS should be regarded as just a type of GIS. The reason for this is a desire to emphasize the different types and needs of users. These differences are reflected by the different sorts of data libraries and analytical 'wizards' incorporated within GIS and GDIS. Within the latter there are particular foci on linking datasets to commonly used and understood 'building block' geographies, and on making relative and neighbourhood-based comparisons between some source of geodemographic information and some other baseline, geodemographic dataset. Nevertheless, we recognize that the distinction is not always so clearly cut, particularly when geodemographic software are built within pre-existing GIS architecture. Longley *et al.*'s (1999, p. 1) generic description of GIS as a 'term denoting the use of computers to create and depict digital representations of the Earth's surface' could encompass GDIS, although we give more emphasis to representing the socio-economic

Table 5.7 Summary comparison of geographic and geodemographic information systems

	GIS	GDIS
Primary data types	Raster, vector	Vector
Primary data sources	RS, surveying, GPS, scanning, digitization, photogrammetry, public surveys and census, data warehouses	Public surveys and census, commercial marketing information, data warehouses, client lists
Typical applications	Agriculture, conservation, environmental assessment and management, emergency management, land-cover analysis, land-use monitoring, landscape conservation, military, public-service delivery, utility, telecommunication and transportation management	Retail, marketing and service management
Typical types of analysis	Data overlay, geoprocessing, image analysis, feature extraction, attribute query, 'hot-spot' analysis, geostatistics, 3D modelling	Neighbourhood profiles and comparisons, cross-tabulations, frequency analysis, ranked correlations, catchment analysis
Scales of analysis	Flexible	Hierarchical, defined by the 'building block' geography
Primary types of output	Tabular, charts, maps	Tabular, charts, maps

attributes of populated places than to the physical properties of land-scape.

A final comparison of GIS and GDIS is offered by Table 5.7. In summary we may characterize GDIS as offering bounded flexibility, relative to GIS. By focusing on the geodemographic subset of all geographical data types available, GDIS offer specific flexibility to handle and relate geodemographic information in ways the analyst may find harder to achieve in conventional GIS.

Summary

- Geodemographic information systems (GDIS) are software used to capture, store, transform, analyse and display *geodemographic* data.

- Geodemographic data offer information about people and the places where they live but are usually aggregated into postal or census zones that define a 'building block' geography for analysis.
- A GDIS will be built around a pre-given classification of the neighbourhood zones and may include other local market, consumer and demographic data.
- Linking proprietary data to the neighbourhood classification may first require a period of 'cleaning' the data, for example removing duplicate records and cross-referencing the accuracy of the information against other sources of data.
- Data analysis in a GDIS – including neighbourhood profiling and catchment analysis – is relational; to also be meaningful requires that sensible relationships be specified.
- Interpretation issues can arise when comparing and relating apparent differences between neighbourhoods so it is important not to draw conclusions that are not substantiated by the data.
- The risk of ecological fallacy (in various guises) needs especially to be considered.

Further Reading

- Johnson, M. (1997, originally 1989) The application of geodemographics to retailing: meeting the needs of the catchment, *Journal of The Market Research Society*, **39**, 203–24.
- Openshaw, S. and Blake, M. (1995) Geodemographic segmentation systems for screening health data, *Journal of Epidemiology and Community Health (supplement)*, **49**, 34–8.
- Tickle, M., Brown, P., Blinkhorn, A. and Jenner, T. (2000) Comparing the ability of different area measures of socio-economic status to segment a population according to caries prevalence, *Community Dental Health*, **17**, 138–44.

Plate 1 GIS mapping of the SuperProfiles classification of 1991 Census enumeration districts in and around Bristol, England

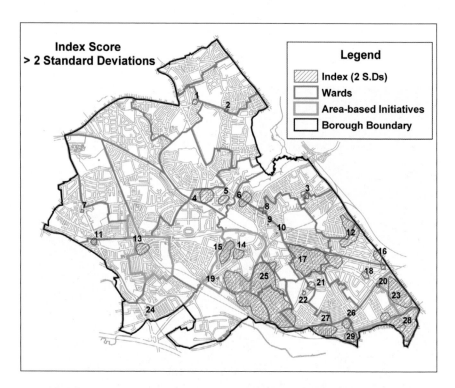

Plate 2 Within Brent, London an index of deprivation shows that changes in geo-demographic condition crossover the administrative ward geography and that not all deprived areas are in neighbourhoods that were receiving regeneration funding. Source: courtesy of Gillian Harper, Birkbeck College

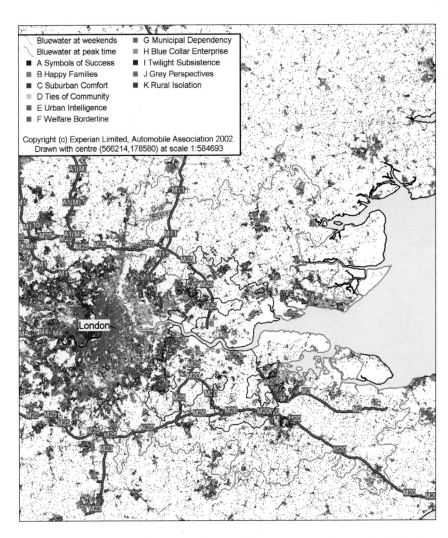

Legend (within map):

Bluewater at weekends
Bluewater at peak time
■ A Symbols of Success
■ B Happy Families
■ C Suburban Comfort
▨ D Ties of Community
■ E Urban Intelligence
■ F Welfare Borderline

■ G Municipal Dependency
■ H Blue Collar Enterprise
■ I Twilight Subsistence
■ J Grey Perspectives
■ K Rural Isolation

Copyright (c) Experian Limited, Automobile Association 2002.
Drawn with centre (566214,178580) at scale 1:584693

Plate 3 Point map showing the catchment of Bluewater shopping centre defined as postcodes within a 45-minute drive time at (i) peak, weekday periods and (ii) off-peak at weekends

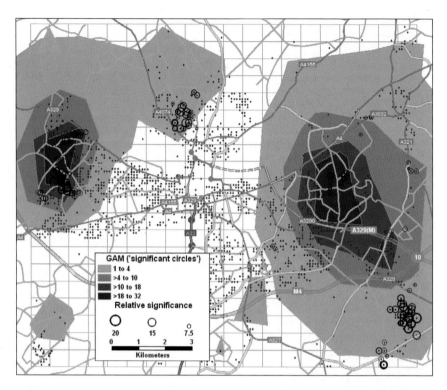

Plate 4 Targeting clusters of 'singles' – comparing GAM with the 'hot-spot analysis' shown in Figure 9.6

6
How Geodemographic Classifications are Built

Learning Objectives

In this chapter we will:

- Look at the techniques used by one company, Experian, to build geodemographic classifications.

- Review the sources of data used, how they are prepared for classification, the role of manual intervention in the clustering process, the formation of a cluster hierarchy and the production of supporting material to describe, represent and help users make sense of the clusters.

- Consider the need for data standardization, variable weighting and the avoidance of small sample sizes, and the methods used to achieve these.

- Consider how minimum spanning trees, principal components analysis and variable weighting can be used to identify and correct for cross-correlations within the variable set.

- Outline some of the advantages and disadvantages of two clustering techniques – the stepwise, top-down approach and the iterative allocation-reallocation method (*K*-means).

- Present a worked example of the *K*-means method.

Geodemographics, GIS and Neighbourhood Targeting   Richard Harris, Peter Sleight and Richard Webber
© 2005 John Wiley & Sons, Ltd   ISBNs: 0-470-86413-3 (HB); 0-470-86414-1 (PB)

Introduction

In this chapter we consider the techniques used to build geodemographic classifications. When scientists develop methods for classifying geological series, climatic zones or vegetation cover, they build classification systems that cross geopolitical boundaries. By contrast, while commonalities can be identified for types of neighbourhood in different countries (and we do so in Chapter 7), there are many practical reasons why it is better for geodemographic classifications at first to be optimized on a country-by-county basis. These reasons include the fact that no two countries have the same data infrastructure; that different sources of data are available for different countries; that these data are available at various levels of geographic detail; because regulations governing access to data also vary; because countries do not all have the same data update frequency; and, while there is increasing harmonization on the questions covered by national censuses, no two census agencies make available the same (identical) set of variables for use in building classifications.

Despite these variations, the methods used by different companies to build classifications in different countries have many similarities. Companies, including Claritas, CACI and EuroDirect (see Chapter 3) use statistical methods for building classifications that have many elements in common with those described in this chapter but will seek to use exclusive sources of data that make their classifications unique. Here we look at the sources of data used by Experian, how they are prepared for classification and then evaluated, how weights are selected, the clustering method employed and how the process is 'optimized', the sorting of the clusters into a hierarchy and their labelling with a 'pen portrait'. In addition, we give a worked example of how a dataset can be clustered.

6.1 Data input – sources of data for neighbourhood classification

We have already commented that no two countries' data infrastructure is the same; the developer can only work with what there is available! In many countries, such as China, Hong Kong and Peru, the census is the sole source of data used to build geodemographic classifications. However, in many other countries the census is supplemented with statistics from other sources. Examples of these other sources are electoral registers (UK, Australia, Spain), the files of mail order companies (Netherlands), car registration files

(Italy), property registers (Germany, New Zealand, UK), registers of shareholders and of directors (UK), statistics on house prices and on council tax bands (UK), and registers of addresses (Australia, UK). In the Netherlands, where census statistics are not published at a small-area level, market research respondent files are used. In the UK the results from lifestyle questionnaires (see Chapter 9) can be incorporated into the build.

These non-census sources of information may be useful for a number of reasons. Understandably, questions in national censuses tend more strongly to focus on measures of disadvantage than on measures of affluence, asking their populations about their literacy (as in Brazil, China), long-term illness (UK) or unemployment (in Australia, for example). Information from non-census sources, such as director or shareholder registers, often is helpful in redressing the balance and in providing greater detail about the location of more privileged members of the community.

A second advantage of using non-census sources is that in many instances it is available at a finer level of geographic detail than that at which census statistics are published. A third advantage is that in many markets the use of non-census sources makes it possible to update the classification codes given to existing areas as their population profile changes over the 5- or 10-year interval between most censuses. Likewise, by using non-census data sources it is possible to assign classification codes to neighbourhoods built since the date of the last census.

While originally Experian were wary of introducing variables that were not demographic in nature into the classification process, in the 1991 Great Britain (excludes Northern Ireland) and more recently in the 2001 Australian and UK classifications, increased emphasis has been placed on the use of statistics that measure the accessibility of zones to shopping centres, proximity to the ocean and generalized measures of closeness to other populated areas.

Where the census is the sole source of information used to classify neighbourhoods then the classification will be built at whatever level of geography the census authorities use for the publication of their smallest area statistics. However, where information used originates from multiple sources, the unit of classification may be more detailed than that of the census. So, in the UK, for example, where there are usually about 5–10 postcodes for each census output area (OA), different postcodes within a single OA may be assigned to different types of neighbourhood.

A problem with focusing on the minutiae, though, is that it loses sight of the bigger picture. For example, areal units with apparently similar census profiles may nevertheless be characterized by quite different social processes, opportunities or economic circumstances according to

Table 6.1 Standardized index values for two postcodes using census variables first at the OA scale and then aggregated using a 5 km buffer zone

Aggregate census index	OA		5 km	
	A	B	A	B
Professionals	122	138	50	99
Agricultural	128	0	1846	39

Note: An index value of 100 indicates the national average (50 is half, 200 is twice the national average).
Source: Adapted from Howick (2004).

their separate locations and regional contexts. Since 2003, Experian now incorporate into each new Mosaic build from the census, which are calculated for a series of concentric circles drawn outwards from each small area zone and including more and more of the regional context with each step out. The purpose of this innovation was to differentiate suburban areas in very small service towns serving agricultural hinterlands from areas of similar demographics located in the inner areas of larger metropolitan suburbs.

Table 6.1 shows, for two postcodes (A and B) in England, two such variables, standardized against a UK index value of 100 as the mean average for each variable. Looking at the relative proportion of professional persons in each postcode it is found that when represented at the OA scale, postcodes A and B appear similar (A has an above average index value of 122, B has 138). However, if all OAs within a buffer zone of 5 km from each postcode are considered, then A is shown to have half the average proportion of the population classified as professional, whereas B is almost average. Looking at the proportions of agricultural workers it should be clear which of the two postcodes lies within an agricultural hinterland! The concentric circle method (not to be confused with Burgess' model, see Chapter 2) has also proved to be very efficient in improving the ability of the classifications systems to predict variations in the level of factors such as risk of crime (cf. case study Chapter 5, Section 5.3).

6.2 Preparing the data for classification

When the information gathered for the classification system is not consistent in terms of its geographical resolution, it is necessary to link together the data for these different levels into a single 'rectangular' database, spreading data for larger geographical units across all the lower level geographic units that fall within them. Building the relationship between the larger and the smaller units can often take some time, especially where the postcode system used to report non-census data sources does not mesh with the administrative geography in terms of which census statistics usually are reported. One key advance in the design of the 2001 UK Census is that the areas for which census statistics are published generally now consist of combinations of whole postcodes (see Chapter 8, Section 8.1.1).

Assuming statistics have been sourced from available databases and that they have been linked together into a single geography, the next stage in the process of building a classification is to create a set of variables as input into the clustering algorithm that is used to build the classification system. As a general rule, but within limits, the more variables that are used in the clustering algorithm and the more different sources they come from the more meaningful (nuanced not idiosyncratic) the resulting set of clusters is likely to be. However, it is important that variables should be included in the clustering process only if they can be seen to be reliable, robust and as adding new information (not repeating what is already known). The evaluation and weighting process described below therefore is an important stage in the build process.

Very often the sourced data items will arrive in the form of counts, such as total numbers of cars. Each of these counts needs to be related to a corresponding base count. For example, total cars in a given zone (the numerator) could be related to the base count of total population, to the base count of total adults or to the base count of total households (each could serve as the denominator). Alternatively the count could be used against all three base counts to create three separate variables for use in the classification system (notwithstanding the issues of correlation and duplicating information raised above).

The classification builder will also spend time at this stage deciding how to group some of the counts. For example there may be available from the census the number of residents aged 0–4, 5–9, 10–14, 15–17, 18, 19–20 and so on. Should each of these separate counts be divided by total residents to create variables for each of these individual age groups? Or, would the number of persons aged 18 be too small to create a variable in its own right? Would it be better to form fewer more robust indicators, for example 0–4, 5–9, 10–14, 15–20? Should even coarser bands be created? Alternatively should the clustering process use both finer five-year bands as well as coarser 20-year bands? Likewise when we examine the statistics on employment by industry, should construction be included as a variable in its own right or should it be grouped with energy and transportation into a less specific but statistically more robust variable? Decisions of this sort need to be informed by the statistical reliability of the possible variables and by knowledge of whether, by merging adjacent classes, the detail that is lost is significant.

In general terms the classification builder, when in doubt, is likely to create a number of alternative measurements of any particular topic and to develop a set of strategies for evaluating which of these alternatives are likely to be most appropriate for inclusion in the classification process.

6.3 Evaluation of input variables

In this stage of the process the classification builder will apply various strategies for evaluating the appropriateness for including different variables as the basis for clustering.

It may be that some variables, for whatever reason, are not deemed appropriate for use in the clustering process. Of those that are to be included, an equally important outcome of the evaluation process is a decision on the weight that should be given to each variable. Consider this by analogy to a recipe, which sets out not only the ingredients to be included in a stew but also the relative quantities of the ingredients that should be used (e.g. the gristle to dumplings ratio). Likewise with the cluster algorithm, the choice exists as to how much emphasis is to be placed on each of the different input variables when calculating, for each zone, the cluster than it is statistically 'closest' to.

6.3.1 Standardization and the problem of skew

One important consideration is the extent to which a variable is skewed. In an ideal world we would include as clustering variables only those which have a bell curved, normal (or Gaussian) distribution. In practice many important dimensions that need to be included in a classification are not normally distributed. The residential location of people who work for the military tends very tightly to be concentrated in a small number of garrison locations. Newly arrived immigrant ethnic groups tend to cluster into particular localities (indeed, this was the basis for the Burgess model in Chapter 2). People who work in agriculture tend also to be concentrated in a limited number of census OAs. These groups of people are not normally distributed across geographical space.

Transforming input variables involves subtracting, from the zonal value for each variable (z_j), the national mean for that variable (μ) and then dividing the result by the national standard deviation (σ), which is the square root of the national variance around the mean – a measure of the diversity between areas in terms of how much of the variable they have:

$$z_j^* = \frac{z_j - \mu}{\sigma} \tag{6.1}$$

where

$$\sigma^2 = \frac{\displaystyle\sum_{j=1}^{N}(z_j - \mu)^2}{N} \tag{6.2}$$

and

$$\mu = \frac{\displaystyle\sum_{j=1}^{N} z_j}{N} \tag{6.3}$$

N is the number of zones (e.g. census or ZIP code areas) to be clustered.

The result of the transformation is to express each zonal value in terms of its standard deviation from the mean for the variable. The resulting value is called a 'z-score' (unfortunately this risks confusion with the notation adopted in some GIS texts where z generally indicates an attribute value). The reason for producing z scores is to allow comparison across a range of variables of how different from average a particular area is. The standardization is required because whereas a shift of, say, 10 percentage points from the mean may not seem overly unusual with regard to a variable exhibiting considerable variance between areas, the same percentage shift for a variable exhibiting less variability could be unusual. For example, if three-quarters of all observations have a value between 30 and 70, then the fact that one particular area has a value of 40 is hardly unusual. However, if three-quarters of observations were in the narrower range 45–55 (suggesting less variance) then a value of 40 is more surprising and would result in a higher z-score.

A problem with comparing sets of z-scores for different input variables is the assumption that the distribution of each set is symmetrical about the mean and has the same shape for each variable, otherwise like is not compared with like. Assessment of the statistical significance of any given z-score conventionally also requires the distributions to be normal. A skewed (non-normal, non-uniform) distribution violates these assumptions. When an extreme outlier on a highly skewed distribution exists for a zone it will be given a z value of very large magnitude, which could have a disproportionate effect on determining that area's cluster membership. This is clearly unsatisfactory and various strategies can be used to avoid it happening. For example, the variable can be transformed using a log function. Alternatively, an upper limit can be applied to the values, with all values exceeding the threshold being reduced to it. Both these methods are quite appropriate for variables such as population density or distance from the coast. However, for many demographic variables Experian classification builders normally refrain from using these methods, preferring instead to reduce the weight given to much-skewed variables to levels at which extreme scores do not override all other criteria when assigning zones to their best-fit clusters.

6.3.2 Avoiding small sample size

Another important consideration is the extent to which variables have adequate samples. Suppose we had access from the (UK) census to the number of people by OA aged over 100. Although this might at first glance be a very interesting variable to include in the classification, we have to bear in mind that there is likely on average to be only one person aged over 100 in every 20 census OAs. When we consider that perhaps 95% of the people who were over 100 on census night, April 2001, are likely to have died by the time the census results are published, September 2003, it is evident that this variable, if used in a classification, is not going to be particularly useful for describing a census output area in 2004 let alone in 2010, the year before the next census!

Small sample size also is a particular issue when variables are sourced from files which are not universal in their coverage – for example, market research files in the Netherlands and lifestyle surveys in the UK. In these instances it is best to use variables which combine a large mean (such as people who smoke) and which have a large standard deviation at the OA level (such as people who have a garden or live in a house built before 1945). One strategy for using these variables may be to avoid overconfidence in their geographic detail. For example, though the proportions of smokers may be calculated at a postcode level the sample size is small and it may be safer to use the variable only at the much coarser OA level (cf. Chapter 9, Section 9.1).

The variable 'aged over 100' is a good example not just of a variable which has too small an average value for the unit of geography for which it is calculated but also of a variable with an unstable value over time. Given the longevity of a classification, it makes sense, other things being equal, to select variables whose values at a small area level change only slowly over time (such as the percentage of households living in apartments, the percentage of buildings constructed before 1945). It makes less sense to use variables whose rank order of zones is likely to be volatile from year to year. Examples of the latter are variables relating to change itself, such as the number of houses sold in the previous year, the proportion of households who have moved, changes in the level of house prices, unemployment and so forth. High values on any one of these variables in any one year are seldom reliable indicators of continuing high values at any subsequent point in time.

6.3.3 Can the variables be updated?

Another important consideration when evaluating the suitability of candidate variables is whether or not they reliably can be updated over time. The importance of census variables to the classification process is that they tend to have a high level of reliability and a near complete national coverage. The disadvantage of census variables is that they are updated only as often as the census is, which in most countries is only once in 10 years. While non-census data sources are seldom as robust and reliable as the census, updatability may justify giving them a greater weight in the classification process than otherwise would be the case.

6.3.4 Variable cross-correlation and the minimum spanning tree

Once the various measures have been evaluated in this way the candidate variables are then correlated with each other and the results expressed in the form of a minimum spanning tree, as in Figure 6.1. The value of the minimum spanning tree, or single linkage analysis as it is often called, is that it identifies sets of variables which have particularly high correlations with each other, whether positive or negative. The tree will therefore highlight 'duplicate' or nearly duplicated variables. In Figure 6.1 each variable has been linked to the other variable in the dataset with which it has highest correlation, whether positive or negative. The thickness of the lines indicates the strength of the correlation. Care is needed when interpreting the minimum spanning tree. Correlations are *positive* between pairs of variables when *either* both of them have a '+' (when the value of one variable goes up so does the other) *or* when both of them have a '−' (when one goes down, so does the other; this is the same as saying both rise together). Otherwise the correlations are negative (as one goes up the other goes down and vice versa).

The classification builder may wish to choose from among duplicate variables the one(s) which have links with the larger number of other variables, signifying a higher level of correlation – this being an indication of the extent to which the variable is likely to be reliable and robust. Alternatively more than one near duplicate variable may be included in the classification process but with each one being given a lower weight than would otherwise have been the case.

The ability of the minimum spanning tree to group variables into 'islands' can also be helpful for classification builders since it will alert them to factors or domains which are under or over represented in the dataset. Classification builders may want to ensure that rural areas are clearly identified by the classification. Taking 'rurality' as a domain we see from Figure 6.1 that it is covered by variables from a number of different

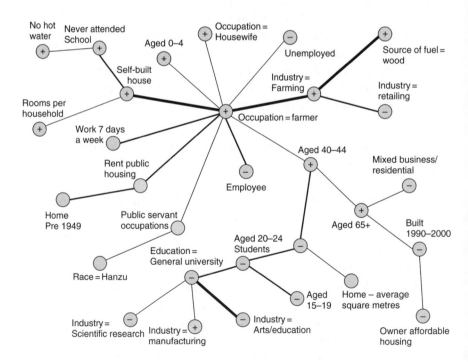

Figure 6.1 Minimum spanning tree of variables for three cities in China

topics in the census, for instance by occupation (farmer), by industry (farming) by employment status (housewife) and by mode of heating (wood). Knowing the extent to which the domain is covered by the variable set is important when decisions are taken about the weight given to individual variables.

6.3.5 Principal components analysis

One approach that is not used in the Experian methodology is principal components analysis (PCA) or the related technique of factor analysis. These are methods used to isolate the main differentiating factors or components of a group of correlated variables (Robinson, 1998). To use the example above, the factor common to 'farmer', 'farming' and 'wood' is rurality. Experian's experience has been that use of these methods tends to blur rather than clarify fine distinctions between cluster types. However, PCA has been applied when building some geodemographic typologies – for example, the post-1981 UK Census version of the SuperProfiles classification. The rest of this section describes how PCA operates. It is a little technical and can, if desired, be understood as saying that PCA trims the initial dataset, removing redundant but perhaps also important information. With that knowledge you can safely skip to the next section!

To build the post-1981 SuperProfiles classification, Charlton, Openshaw and Wymer (1985) report that a preliminary selection of 465 census variables was made. These variables were grouped together according to their similarities with each other and those groups refined to leave 55 cluster formative variables. PCA was then applied to those remaining variables. The first stage of the PCA was to calculate the inter-correlation or r-matrix of m ($=55$) cluster formative variables across the N ($=130,000$) zones to be clustered. The correlations are in the range from -1 to $+1$.

The second stage of a PCA procedure is to transform the raw data matrix by the inverse cosine of each correlation found at the first stage. Describing the geometry of PCA in detail, Robinson (1998, pp. 121–3) notes that this stage is equivalent to calculating an angle between each pair of variables. These angles will be in the range from $0°$ to $180°$ since cosine^{-1} of $r = +1.0$ is $0°$ and cosine^{-1} of $r = -1.0$ is $180°$. Furthermore, each variable can now be represented as a vector, each of length $|d|$ from a fixed origin (O), where the angle between each pair of variables is defined by the transformed correlation matrix. Since vector quantities are analogous to forces, since each variable (or 'force') pulls upon the origin in different directions and since each force has constant magnitude (defined as $|d|$ above), so the resultant force can be found. This is the combined effect of each of the individual forces (or variables) and gives the direction the origin would be forced to move were it not fixed. The resultant is called an eigenvector and is known as the first principal component.

The principal component eigenvector, like the input variables, can be described as a force directed out from the same fixed origin, O. By extension, the angle between each variable and the principal component may be calculated ($\theta_1 \ldots \theta_m$). The smaller the angle, the more it may be said that the variable and the first component are 'pulling in the same direction'. A new correlation matrix giving the correlation between each variable and the principal component is derived as the cosine of the angle between them, θ (this is the inverse of the transformation undertaken in the second stage). A perfectly correlated variable has $\theta = 0°$ and thus $r = +1.0$.

Each correlation defines the loading of a variable on the component. The variance among the variables that is explained by the component is equal to the sum of the squares of the loadings across the m variables. The first component will always explain the most variance, while each subsequent component will explain less than the component which preceded it. The second component is erected at a right angle to the first and each consecutive component is also arranged in the same way (orthogonally). The procedure can be continued until there are as many components as there are input variables (m). However, since PCA is intended to eliminate duplicated information within the variable set so the actual number of components selected will be that sufficient to explain a certain threshold of variance among the variables.

For the SuperProfiles classification, 27 components (based on 55 input variables) were needed to account for over 90% of the variance of the initial correlation matrix. Curry (1993, p. 204) considers these types of procedure more generally, arguing that it is largely redundant information that is removed and that upwards of 80% of the valid or non-redundant information contained in the original data matrix will be retained in the trimmed matrix. Of course, the same figure could also be quoted as a loss of 20% of the valid data – perhaps too much, especially if it is the fine distinctions between cluster types that are lost.

The last stage of PCA is to replace the initial $m \times N$ data matrix (the input variables) with the smaller $q \times N$ matrix, where q indicates the chosen quantity of components ($q < N$). Assuming the original data are expressed in terms of their standard, or z-scores per variable (see

Equation 6.1, above) then the component score, c_{ju}, of the uth component for the jth area is:

$$c_{ju} = \sum_{v=1}^{m} z_{jv}^{*} l_{uv} \qquad (6.4)$$

where l_{uv} is the component loading of the vth variable on component u and z_{jv}^{*} is the standardized value of the vth variable for area j. A complete matrix of 'reduced data' is obtained by calculating each of the component scores in the same way for the total of N zones and q components (cf. Robinson, 1998).

Aside from the potential problem of information loss, a difficulty with PCA is making sense of the outcome! This is not necessarily a problem for geodemographic classifications that feed the trimmed matrix into the subsequent cluster analysis and which then seek to explain the characteristics of the clusters by cross-referencing back to the 'raw' (and other) data – see the discussion on labelling, Section 6.8 below. However, there are cognitive disadvantages when working with a dataset that is no longer 'intuitive' – particularly when the whole nature of cluster analysis is as much an art as a science.

6.4 Selecting weights

Instead of using PCA, Experian's strategy is to give careful consideration to the appropriateness of the weights given to different variables in the clustering procedure. Assigning weights has the effect of reducing (or increasing) the influence of variables which, from the minimum spanning tree, are known to share information (or not) with other variables.

As we have seen, there are many considerations that need to be taken into account when setting the weights. A useful check on the appropriateness of the weighting strategy is to calculate the share of the total weight that is accorded to variables belonging to different 'domains'. In the Australian classification, fixed proportions of the total weight were assigned in advance to the sets of variables pertaining to: housing; social and economic status; age; household composition and cultural identity; and, finally, accessibility. Likewise, decisions were made as to the relative weight to be accorded to the three levels of geography used in the classification: street segments, census collection districts and concentric circles. The weight initially given to every variable after the evaluation process was then adjusted to ensure appropriate overall weights for these domains and geographic levels.

Table 6.2 Weights assigned to age variables when building UK Mosaic

Variable	Weight	Variable	Weight	Variable	Weight
Aged 0–4	20	Aged 30–34	40	Aged 60–64	40
Aged 5–9	40	Aged 35–39	40	Aged 65+	40
Aged 10–14	40	Aged 40–44	40	Aged 65–74	4
Aged 15–19	40	Aged 45–49	40	Aged 75–84	4
Aged 18–24	10	Aged 45–64	40	Aged 85+	15
Aged 20–24	40	Aged 50–54	40	Aged 85–89	2
Aged 25–29	40	Aged 55–59	40	Aged 90+	5
Aged 25–44	40				

Table 6.2 illustrates the weights given to different age bands in the construction of UK Mosaic. From this table you can see that overlapping class intervals have been selected at the upper bands of the age distribution but also that these have been given lower weight than those allocated to the younger age groups.

6.5 Clustering

Once the variables have been defined, selected and given appropriate weight, the clustering process begins. It is important to understand that the categories of neighbourhood used in a geodemographic classification are not defined in advance. The builder does not start with the requirement to find a category called 'Rural Isolation' or 'Mortgaged Families'. Such categories may or may not emerge from the cluster build. However, from his or her prior experience in different countries, the builder of a geodemographic system will have some expectation that the computer programs will identify a number of neighbourhood types bearing strong similarity to those identified in other markets. It is for this reason that Experian has been able to identify a set of 13 global 'lifestyle groups' ranging from 'Agrarian Heartland' to 'Shack and Shanty' that tend to be found repeatedly across each of the 18 different national markets where its systems operate. Needless to say not every category is found in every market and there are types of neighbourhood that are specific to particular countries. We discuss these categorizations further in Chapter 7.

There are a number of clustering algorithms available but the most common can be broadly grouped into two types: stepwise, top-down methods and iterative allocation-reallocation procedures (two alternatives, simulated annealing and neural network classifiers are described by Openshaw and Wymer, 1995, pp. 255–66).

6.5.1 The stepwise, top-down method

Of the more common types, the stepwise approach conceptually is the simpler. If initially there are N zones then these are each taken to be a 'cluster'. Hence, there are as many clusters as there are zones – too many! To reduce the number of clusters the two that are most alike are grouped together. These are found by considering every possible pair of clusters in turn and determining how alike or close together they are with respect to the set of input variables used for the clustering. Once a pair is identified and merged the values of each of the variables are recalculated for the merged clusters by taking an average of their constituent members.

There are now $K = N - 1$ clusters left to consider at the second step (i.e. one less than previously). Again, the two that are now closest are identified and merged, and the input variables averaged for the group. The process of grouping like-with-like can be repeated until there is only $K = 1$ cluster left (i.e. all the zones are merged into one). In practice the process would stop before this, at the value of K that was felt to achieve the best balance of differentiating *between* neighbourhood types while at the same time avoiding too much heterogeneity (variance) *within* neighbourhood types.

The major disadvantage with the top-down approach is how long it takes to compute. If N zones are finally grouped into K clusters then the total number of cluster pairs that must be checked to see whether they are closest together is actually equal to:

$$\frac{1}{6}[N(N + 1)(N - 1) - K(K - 1)(K - 2)] \qquad (6.5)$$

If this checking involves looking along m cluster variables, then the total number of calculations that must be performed is at least:

$$\frac{m}{6}[N(N + 1)(N - 1) - K(K - 1)(K - 2)] \qquad (6.6)$$

To put the formula into some sort of perspective: if $N = 100,000$ zones are grouped into $K = 10$ clusters on the basis of $m = 100$ input variables then the total number of calculations that must be performed to arrive at the solution is approximately 10^{16} – ten thousand trillion! Worse still, there is no guarantee that it is actually the best cluster solution possible because making the 'best' decision at each particular step does not necessarily lead to an optimal result overall. Consider, for example, a stepwise algorithm at the point that it had created a three-cluster solution. At the next step a two-cluster solution would be created by merging two of the clusters. However, a better solution likely would be achieved if, instead, one of the three clusters were split, with some of its members going to both of the two remaining clusters.

6.5.2 The iterative allocation-reallocation method (*K*-means)

Unlike top-down methods, iterative procedures start with zones imperfectly arranged into the final number of clusters and then repeatedly refine the 'solution' to obtain a better fit. Although there is a computational saving that comes by specifying from the outset how many clusters there should be (thus avoiding the long stepwise process of merging zones/clusters one at a time), the principal benefit of this method is that it generates solutions which retain a higher proportion of the variance of the source variables and clusters which are more equal in population size, particularly where the number of clusters is small in relation to the total number of input zones. To begin, the classification builder has to specify the number of clusters (*K*) that s/he thinks would be appropriate for the market being clustered, bearing in mind the range of data available, the level of geography used and the complexity of the market.

The next stage in the clustering process used by Experian involves the calculation of the means and standard deviations of the input variables, and the standardization of the data. An important feature of this process is that these, and all subsequent, computations are population weighted. That is to say that when calculating the means and standard deviations the algorithm gives correspondingly more attention to the values of zones with high populations than to those with low:

$$z_j^* = \frac{w_j z_j - \mu}{\sigma}$$

(6.7) (cf. (6.1))

where

$$\sigma^2 = \frac{\sum_{j=1}^{N}(w_j z_j - \mu)^2}{\sum_{j=1}^{N} w_j}$$

(6.8) (cf. (6.2))

$$\mu = \frac{\sum\limits_{j=1}^{N} w_j \, z_j}{\sum\limits_{j=1}^{N} w_j} \qquad \text{(6.9) (cf. (6.3))}$$

and w_j is the weight assigned to the jth area and is proportional to that area's population count.

The population weighting particularly is important insofar as the populations of rural and inner city zones will in many markets tend to be lower than the populations of zones containing new housing estates on the edge of urban centres. Without using population weighting the clustering process ignores that fact that some places contain a far greater population – and likely more diversity – than others. It would not be desirable for a cluster solution to have, for example, as many different types of rural neighbourhood as it does urban types if the rural areas contain the much smaller proportion of the total population. In practice it could be appropriate to select numbers of households or adults instead of the entire populace as the basis for weighting.

The next stage of the clustering process involves the selection of a set of seed zones. If the classification builder has required the cluster algorithm to create a 45-cluster solution then 45 zones in the dataset will at this stage be selected to form the initial nucleus or centre of each cluster. The zones will be selected on a fixed interval basis with a probability proportionate to their population. The selection of the first zone occurs once half of the sampling interval is reached. The method of selection is quasi-random in that the zones are not pre-sorted on any particular variable. They do not enter in any systematic order other than by county so there is likely to be some regional stratification in the selection of seeds.

The algorithm will now examine every zone in the database and, taking into account the standardized data and the weight assigned to each variable, calculate a measure of similarity between that zone and each of the 45 seeds. This measure is known as the k-mean squared distance (because k means or centres have been selected as the basis for clustering). It is calculated by taking the squares of the differences between the standardized scores of the zone and the standardized scores of the seed zone, summing across all variables and weighting by variable:

$$d^2 = \sum_{v=1}^{m} w_v (z_{jv}^* - z_{kv}^*)^2 \qquad (6.10)$$

163

where m is the total number of input variables, z_{jv}^* is the value of the vth variable for area j, z_{kv}^* is the corresponding value of the same variable for the kth seed, and w_v is the weight assigned to the variable. The zone is then assigned to the seed for which this measure of distance is lowest, in other words the zone to which it is most similar.

Note that if there were only two variables then:

$$d^2 = w_1(z_{j1}^* - z_{k1}^*)^2 + w_2(z_{j2}^* - z_{k2}^*)^2 \qquad (6.11)$$

or

$$d = \sqrt{w_1(z_{j1}^* - z_{k1}^*)^2 + w_2(z_{j2}^* - z_{k2}^*)^2} \qquad (6.12)$$

Comparing Equation (6.12) to the measure of distance expressed in Chapter 4, Equation (4.6) reveals that the k-mean squared distance is an expression of the Euclidean distance between zones and seeds, and is only one of a number of distance measures that might be used (see Berry and Linoff, 1997).

Once the allocation process has been repeated across all zones, the situation will arise when each zone is assigned to its nearest or 'best-fit' cluster. The program will then take all the zones which were found to be closer to seed one than to any other seed and calculate the average value of this set of zones on each of the input variables used in the clustering process. This set of calculations is repeated for all the other 44 seeds.

At this point the algorithm will commence a second loop (or iteration) during which it reviews for each of the zones in the database which of the 45 seeds it is now closest to in terms of similarity. For many of the zones the result of this computation will be the same as it was on the previous iteration. However, since the averages of the seed zones have now been replaced by the averages of the zones assigned to them after the first iteration, a number of zones will now find themselves shifting from the cluster to which they were first assigned to a different cluster.

At the end of this iteration the averages are recalculated for each of the 45 seed clusters and a further iteration is done. This process continues, often for more than 20 iterations. However, each additional iteration will usually lead to a smaller number of zones moving from one cluster to another (compared to the last) and also progressively smaller changes in the values of the variable averages for each cluster. Eventually, when an iteration generates no further change, the process comes to an end. The algorithm has reached a local optimum beyond which it cannot improve and is said to have converged. The stages of the allocation–reallocation procedure are summarized by Figure 6.2. As the final solution is in part dependent on the initial seed selection so it is sensible to either repeat the process to see if different seed selections produce improved results and/or make manual interventions as described below.

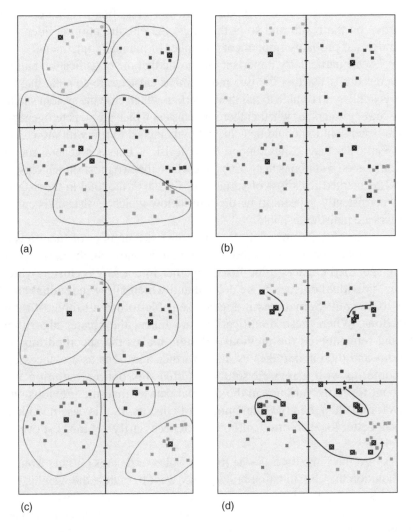

(a)

(b)

(c)

(d)

Figure 6.2 Example of an allocation–reallocation classification with $K = 5$ means. (a) The initial allocation of areas to seeds. (b) The clusters after three iterations. (c) The clusters upon convergence. (d) Movement of centroids during classification. Source: Cameron (2004)

165

6.6 Optimization process and manual intervention

Once the cluster algorithm has found the local optimum it orders the resulting clusters such that clusters that broadly are similar are given consecutive numbering. It then reports to the cluster builder a number of diagnostics from which s/he can assess the effectiveness of the solution. It is likely that by skilful modification the classification builder can improve on the result and so at this stage the classification builder will consider and probably implement a number of manual adjustments.

One particularly important diagnostic that the classification builder will consider is whether the two most similar clusters are so alike that the user is unlikely to be able to tell them apart, in which case the system should have fewer clusters; or whether there are clusters which are so heterogeneous that a better solution could be achieved by being set to run with more clusters. Particular diagnostics support these decisions but, as a general rule of thumb, a good solution is likely to be one where the two most similar clusters could be merged for a loss of variance of 0.22% of the total in the dataset. This consistently appears to be the limit below which further divisions of clusters are indistinguishable.

The classification builder will now use a number of functions to try to improve the classification. These include ones which cause two or more specified clusters to be fused together into a single cluster, which allow individual clusters to be deleted and which allow individual clusters to be split. Sometimes it appears sensible to use all three of these functions. When these modifications are made, the cluster algorithm, having remembered the previous solution, makes the specified modifications and then undertakes as many further iterations as are necessary to come up with a revised solution. Often this optimal solution will account for more variance in the original dataset than the previous one, even for no change in the total number of clusters. In this case it is likely to be a better local solution, although not necessarily yet the best overall solution.

At the same time as s/he merges, splits or deletes clusters from an old solution the classification builder may want to review the weights that were assigned to the different variables. If, for example, the clusters appear to be over influenced by population density to the exclusion of other important differences, the classification builder may want to reduce the weight given to that variable and re-run from the old solution but with new weights. Alternatively the classification builder might want to improve definition in particular areas of the classification, for example by splitting the most high status of the clusters into two, to assist marketers, or by splitting specific ethnic minority clusters, in order to assist in deprivation studies.

There are circumstances, for example in the identification of different types of rural area, where it may be appropriate on completion of the area classification to actually remove the rural areas altogether from the 'solution' and build a new bespoke classification solely for the rural zones. Such a bespoke solution would perhaps use a different number of rural clusters than that identified at the national level and would almost certainly apply a different set of variable weights. The separate classification can then be linked back into the main classification. This bespoke method is particularly effective for dealing with clusters that are picked out by highly skewed variables, as rural clusters often tend to be. By applying this method it has proved possible in the UK to avoid rural classifications being differentiated solely on the basis of level of rurality (which otherwise happens) and instead introduce qualitatively different types of rural community, such as nucleated villages supporting arable farming, as opposed to areas of self-employed dairy and cattle farmers more often living in scattered farmsteads.

Typically it may take five to a dozen different runs of the cluster algorithm before an ideal solution is finally agreed and much time can be spent evaluating which of the alternative solutions is the best one. One method of evaluating the solutions is to see which one explains the most (weighted) variance in the input variables. All things being equal this is a sensible approach. However, it is not an entirely fair adjudication procedure if the number of clusters has been changed or if there have been alterations in the variable weights between the two solutions. A fairer evaluation is to compare the extent to which alternative classifications are effective in predicting variations in behaviour on datasets that were not used to build the classification. Tests against 100 external files enabled Experian to improve the efficiency of their 2001 census-based UK Mosaic by 3% compared with what it otherwise would have been. This process of evaluation is particularly effective in determining the most appropriate set of weights to give to variables in different domains and, perhaps even more important, the set of weights to give to variables at different levels of geographic aggregation.

Other important evaluation methods used at this stage of the build process involve the mapping of the clusters in towns familiar to the classification builder. Likewise, when building UK Mosaic, Experian undertook a photographic tour of the UK, which generated over a thousand pictures of different postcodes. Alternative solutions were then partly evaluated with reference to whether the new cluster allocation seemed to provide a better representation of the photographed postcodes than did the previous cluster allocation.

6.7 Forming a cluster hierarchy

In relatively small and homogeneous countries, such as Ireland, Hong Kong or Peru, a classification will be able to identify around 25 to 30 distinct cluster types. In larger and more heterogeneous countries, such as the USA and the UK, it is possible to recognize a larger number of types, as many as 60 to 65. However, because it is difficult for users to remember each of these types individually, the more detailed neighbourhood 'types' (as they are called) are usually classed hierarchically into a smaller number of neighbourhood 'groups', typically ranging from 7 to 12.

Once the solution is finally signed off at the cluster-type level, attention shifts to the process of arranging the clusters into groups. This process initially is undertaken using a stepwise algorithm, which is integrated within the main cluster program. This process starts by considering which pair of clusters could be merged together while causing least loss of variability in the original dataset. The pair of clusters that could best be merged in this way is likely to be a pair that is very similar in terms of their average scores on the input variables. They are also likely to be ones with relatively small populations. Table 6.3 illustrates how this process operated in the build of a Mosaic for three city regions in China. This is a good classification in so far as the first pair of clusters to merge – clusters 28 and 29 – do so for a loss of variance of only 0.22%. The line after step 27 indicates a seeming step change in the loss of variance for subsequent mergers, implying that this is a good point at which to stop, setting the number of groups at seven.

The stepwise process can be used both to arrange cluster types into a chosen number of cluster groups but also to identify from a set of clusters (either groups or types) which are the most similar, permitting the cluster groups or the cluster types to be numbered consecutively and on an ordinal (rank-ordered) scale. There may, however, be other important considerations that the automated procedure alone does not adequately address. For example, one might want to ensure that the percentage of the population in each of the groups exceeded a threshold of 4% while not exceeding 20% and one might reasonably wish to ensure that all groups

Table 6.3 Stepwise fusion process for three cities within China

Step	Fusing cluster	To cluster	% loss in variance	Step	Fusing cluster	To cluster	% loss in variance	Step	Fusing cluster	To cluster	% loss in variance
1	28	29	0.22	12	23–24	25	0.53	23	6–7	8–12	0.98
2	4	5	0.25	13	6	7	0.58	24	33	34	1.26
3	26	27	0.29	14	17	18	0.59	25	23–27	28–29	1.57
4	2	3	0.30	15	30	31	0.60	26	6–12	13–14	1.74
5	13	14	0.36	16	15	16	0.64	27	17–19	20–22	1.90
6	10	11	0.37	17	20–21	22	0.64	28	30–32	33–34	2.60
7	20	21	0.43	18	23–25	26–27	0.68	29	15–16	17–22	3.46
8	23	24	0.44	19	8–9	10–12	0.73	30	15–22	23–29	4.70
9	8	9	0.47	20	1–3	4–5	0.75	31	1–5	6–14	5.29
10	10–11	12	0.50	21	17–18	19	0.88	32	15–29	30–34	7.47
11	1	2–3	0.51	22	30–31	32	0.88	33	1–14	15–34	21.22

contained more than one cluster but not more than seven – this being the maximum number of subcategories, which on an empirical basis it seems practical to distinguish between at any one time.

This process of manual intervention is facilitated by the drawing of a 'family tree'. This diagram, which in structure is not unlike the minimum spanning tree of variables, links each cluster to another cluster which it is most similar to within the set. However, unlike the situation with the stepwise process, linked clusters are not combined only associated; within each linked pair of clusters the tree will identify the one (of the two) which has the highest degree of similarity with a cluster belonging to another grouping. Figure 6.3 shows a minimum spanning tree of clusters for UK Mosaic. Note that each cluster is coloured according to the group that it belongs to. Note also indicators around the perimeter of the tree showing the polarities both in terms of demographics and marketing orientation. This device has consistently proved helpful in identifying the major dimensions of differentiation within the system and these are often set out around the outside of the tree to orient the user. The final decision on how the clusters should be best organized into groups is seldom undertaken without some reference to such a diagram.

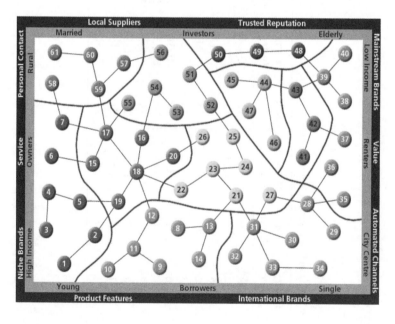

Figure 6.3 Minimum spanning tree of clusters for UK Mosaic. Source: Reproduced by permission of Experian UK

6.8 Labels, portraits and visualization tools

When the clusters have been arranged into groups it is necessary to undertake a final re-ordering in which it is customary to assign the category 'one' to the cluster with the highest status, whether in terms of income, social class or value of housing. Next begins the complex process of assigning 'labels', typically of up to 20 characters, to each category. These labels will then be supplemented by 'portraits', each portrait being a comprehensive account of the demographics and behavioural features that make each cluster distinctive. In some markets these portraits will run to an average of 200 words. In better-developed markets such as the UK typically they average 1000 words. While it is usually the portraits rather than the labels that are used in analysis and targeting, the labels are understandably more convenient for day-to-day communication among teams of people already familiar with a particular classification.

The labelling process is often contentious if only because the expectations and requirements of marketers, public sector users and academics can differ. Marketers typically want to communicate among each other using labels (such as 'Caring Professionals'), which are both recognizable and memorable. The more the label provides insight into the perceived mindset of residents of the cluster, the better. Public sectors users are more mindful of the political correctness of the labels and associated issues of stereotyping, bearing in mind that geodemographics may be used in reports to elected representatives. They would ideally like 'labels' which focused on demographics and which were rather more 'matter of fact'.

Academics tend to seek research evidence to substantiate the labels used but are often unaware of the more detailed portraits that already exist. When confronted with a label such as 'Low Horizons' they understandably ask on the basis of what evidence (and whose opinion) could such an attribute be selected? The answer to such questions is usually found not just in the set of portraits which accompany the labels but also in the extensive visualization and multimedia material which will support any professionally developed neighbourhood classification (see Chapter 5, Section 5.3). By and large, academic critiques of geodemographics have tended to focus only on the labels. This is not surprising given the labels are written as evocative 'snapshots', designed to catch the eye and usually are prominent in promotional literature. By focusing only on the labels, there is risk of critiques 'down playing' the wider range of portraits and visualizations available and contributing to a misunderstanding of both how geodemographic classifications are used, and also the consultative nature of the relationship between vendors and their clients. Nevertheless, there is room for legitimate debate on the extent and ways by which some applications of marketing and geographical information systems over-simplify the complexity and diversity of human subjects.

Label selection is a very important part of the build process. Good labels are ones which can only be true of the type they are given to, which accurately can be applied to virtually all postcodes within that type, which are memorable and which are not offensive. 'Fledgling Nurseries' is a good example of a label that meets all these criteria. An important rule applied by Experian is that type labels should be no more than 20 characters in length. This precludes the tendency for labels created by committee to become long, bland and lacking in insight.

The more people in the organization involved in the labelling of the categories the more intelligible they are likely to be. However, writing up the descriptions of the categories is a more solitary task. Typically the clusters types and groups will each be subject to a textual portrait. This portrait will look at the clusters from different dimensions: their physical appearance; their historical origins; the types of people who live in them; their values; their consumption patterns; the patterns of movement into and out of the clusters; and ways in which it is likely that they will change over time.

The evidence on which these portraits are created includes a wide range of census and non-census indicators, often a wider range than the restricted set that were used in the build process. Much of this evidence will also typically be included in the accompanying visualization tools. Besides the labels, the portraits and large number of photographs which illustrate the different categories, these tools will typically include libraries of demographic and behavioural variables showing variations between the types and groups in the form of tables and bar charts (see Figures 5.7–5.9 in Chapter 5).

Together with the tables showing demographic and behavioural differences these tools are likely to include charts and maps showing the regions of the country, the local authorities and the constituencies in which they occur. These have proved helpful in enabling users to uncover subtle distinctions between otherwise similar types of places. It is this evidence base that we have relied on in the following chapter to describe some differences in the patterns of residential segregation between Mediterranean and Northern Europe, North and Latin America, and Australia and China. (Multimedia guides for classification systems are often freely available from the developers. For details of Experian's products contact bus.helpdesk@uk.experian.com.) The process of producing the cluster portraits involves a fusion of art and science, of the qualitative and the quantitative and is best undertaken without interruption and with copious supplies of caffeine!

6.9 A worked example of clustering

Imagine you were asked to cluster the hypothetical areas in Table 6.4 into two types, based on the single column of information described as attribute A (ignore the other columns for the time being). The clusters need not be equally sized (in fact, in this case they cannot be) but all areas must belong to one and only one cluster type. Given 30 seconds to find a solution, which areas would you put into which clusters?

Your answer:

Your reasons:

Chances are that you immediately assigned areas a, c, d, e and k into one cluster (call it type I) and areas f, g, h, i and j into the second (call it type II). If so, then to start with, all areas in type I had an attribute A value beginning with a 1 (i.e. in the range 10–19) and all areas in type II had an attribute value beginning with a 3 (in the range 30–39).

This partial solution still leaves area b with an attribute value of 24 to deal with. You were not given the option of not classifying it at all so probably it was assigned to cluster I, given that 24 is closer along the number line to 19 than it is to 30. On this basis you could claim that area b is more similar to the members of the first cluster than it is to the second.

Table 6.4 Area data for worked example

Area ID	Attribute A	Attribute B	Attribute C	z-score A	z-score B	z-score C
a	11	70	2	−1.36	1.51	−1.51
b	24	65	11	−0.09	1.21	−1.21
c	19	60	20	−0.58	0.90	−0.91
d	15	55	29	−0.97	0.60	−0.60
e	13	50	38	−1.16	0.30	−0.30
f	33	45	47	0.79	0.00	0.00
g	36	40	56	1.08	−0.30	0.31
h	37	35	65	1.18	−0.60	0.61
i	34	30	74	0.89	−0.90	0.91
j	35	25	83	0.98	−1.21	1.22
k	17	20	91	−0.77	−1.51	1.48
μ (mean)	25	45	47	0.00	0.00	0.00
σ (sd)	10	17	30	1.00	1.00	1.00

Thinking through the classification process in a little more detail, if you were indeed guided by the fact that nearly half the areas have an attribute value in the range 10–19, and the other half in the range 30–39, then what you may have done (consciously or otherwise) is imagine two cluster centres or averages, the first with a value somewhere in the 10 s, the second with a value somewhere in the 30 s. Each area was then assigned to the cluster with centre closest along the number line to the area's own attribute A value. Therefore, area a, with an attribute value of 11, seems closer to the cluster with its centre in the 10 s than it does to the other. Going on to assign the 'awkward' area b to what we came to call cluster type I simply follows the same logic of looking at the difference or distance along the number line between the centre of each cluster and any particular area's own attribute value. This is illustrated by Figure 6.4.

Things seem more complicated when we introduce a second attribute, B in Table 6.4. However, the rationale remains as simple: the cluster with the closest centre is where areas will be allocated to. The only change is that the centres of our clusters are positioned not along a single

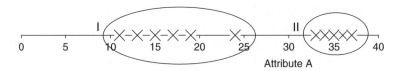

Figure 6.4 Areas are grouped into clusters according to their position along the number line

number line but in a two-dimensional 'space' – one dimension for attribute A, the other for attribute B. Fortunately, two dimensions are still easy to visualize and the data have been plotted in Figure 6.5. It is now easy to perceive to which cluster each area belongs. In this example, the 'difficult' area is k, which is located out at the edge of cluster II.

So far the classification method seems to work. However, difficulties will arise when there are too many areas, too many clusters and/or too many dimensions (attributes) for it to be clearly seen which cluster centre is at least distance to any particular area. Try reproducing something like Figure 6.5 for a set of 100,000 census zones, 10 clusters and 100 variables (actually, don't!). What is needed, therefore, is a way of formalizing the classification procedure so that we are not reliant on grouping the areas by eye. The method we will describe is the *K*-means one.

In addition to attributes A and B, Table 6.4 has a third: attribute C. However, looking back at Figure 6.5 it can already be seen that there will be problems making direct comparisons between these variables: there is less variance (or standard deviation) along the attribute A axis where the spread of values is between 11 and 37, than there is along the attribute B axis where the spread is between 20 and 70. Table 6.4 records that attribute C has even greater standard deviation (σ) and therefore variance around the mean (μ) than either A or B. Consequently, a unit increase in the C value is rather less

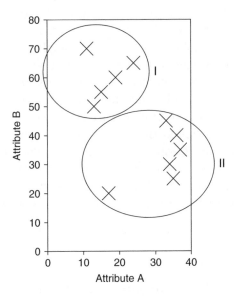

Figure 6.5 Areas are grouped into clusters according to their position within a two-dimensional space

surprising than a unit increase in B which, in turn, is less unusual than a unit increase in A. In order, therefore, to compare like with like, we will now consider the standardized (z-score) values of A, B and C, obtained by using Equation (6.1) and also listed in Table 6.4. For example, the standardized value of attribute A for area a is $(11 - 25) \div 10$ which is -1.36, i.e. almost one and one half standard deviations below the all-area average for attribute A (that it is below is indicated by the minus sign).

To proceed with the clustering we need a starting-point – in fact, not just one but as many as there are clusters sought (K). For illustrative purposes, we will set K to three (one more cluster than before). Previously it was possible to 'see' the centre of each cluster and allocate areas to neighbourhood types accordingly. Now, however, those starting points are not clear. To solve this problem we can either randomly draw three seed areas and take their attribute values as the acting centres of the clusters, or we can instead arbitrarily allocate areas to clusters and then calculate the centres of these arbitrary groupings as the averages of the attribute values per cluster.

To demonstrate how an iterative process of allocation–reallocation gradually improves upon the initial allocations we will use the second option, grouping areas a, b and c into cluster I, areas f and g into cluster II, and areas d, e, h, i, j and k into cluster III. Averaging the attribute values by cluster gives centres at $(-0.68, 1.21, -1.21)$, $(-0.94, -0.15, 0.16)$ and $(0.03, -0.55, 0.55)$, respectively, where (A, B, C) correspond to the three standardized attribute values. The members of each cluster and the average values are shown in Table 6.5. For this example, each of the areas is assumed to be as important as another and so the averages are *not* weighted.

Having established the starting points, areas are removed from their initial cluster groupings and reallocated to clusters with centres closest to their own attribute values. Therefore, each area needs to be compared with each of the three clusters to find the one at least distance from it. That distance is found by comparing pairs of attribute values. Recalling that the temporary centre of the first cluster (I) is at $(-0.68, 1.21, -1.21)$ and knowing, from Table 6.4, that area a is located at $(-1.36, 1.51, -1.51)$, then the square of the distance between area a and cluster I can be found as:

$$d^2 = (-1.36 - -0.68)^2 + (1.51 - 1.21)^2 + (-1.51 - -1.21)^2$$

$$d^2 = 0.64 \qquad\qquad \text{(cf. (6.11))}$$

With the aid of a spreadsheet the distance from every area to every cluster can be found and, with that information, each area assigned to their least distance cluster. Try it! If you get stuck, the answer is shown

Table 6.5 Three steps of an allocation-reallocation clustering procedure ($K = 3$; $N = 11$; $m = 3$)

Step 1 $A_I = -0.68$; $A_{II} = 0.94$; $A_{III} = 0.03$ $B_I = 1.21$; $B_{II} = -0.15$; $B_{III} = -0.55$ $C_I = -1.21$; $C_{II} = 0.16$; $C_{III} = 0.55$

Cluster	Area	A	$(A - A_I)^2$	$(A - A_{II})^2$	$(A - A_{III})^2$	B	$(B - B_I)^2$	$(B - B_{II})^2$	$(B - B_{III})^2$	C	$(C - C_I)^2$	$(C - C_{II})^2$	$(C - C_{III})^2$	Σsq-diff$_I$	Σsq-diff$_{II}$	Σsq-diff$_{III}$	To
I	a	-1.36	0.47	5.27	1.92	1.51	0.09	2.76	4.26	-1.51	0.09	2.77	4.26	0.65	10.79	10.43	I
I	b	-0.09	0.34	1.05	0.01	1.21	0.00	1.85	3.11	-1.21	0.00	1.86	3.11	0.34	4.76	6.23	I
I	c	-0.58	0.01	2.30	0.37	0.90	0.09	1.10	2.11	-0.91	0.09	1.13	2.14	0.19	4.53	4.62	I
	Avg.	-0.68				1.21				-1.21				0.40			
II	f	0.79	2.15	0.02	0.59	0.00	1.46	0.02	0.31	0.00	1.46	0.02	0.31	5.07	0.07	1.20	II
II	g	1.08	3.09	0.02	1.11	-0.30	2.27	0.02	0.06	0.31	2.31	0.02	0.06	7.67	0.07	1.24	II
	Avg.	0.94				-0.15				0.16					0.07		
III	d	-0.97	0.09	3.63	0.99	0.60	0.37	0.56	1.33	-0.60	0.37	0.57	1.33	0.83	4.76	3.65	I
III	e	-1.16	0.23	4.39	1.40	0.30	0.82	0.20	0.73	-0.30	0.83	0.21	0.73	1.88	4.80	2.86	I
III	h	1.18	3.45	0.06	1.33	-0.60	3.26	0.20	0.00	0.61	3.31	0.21	0.00	10.02	0.47	1.34	III
III	i	0.89	2.45	0.00	0.75	-0.90	4.44	0.56	0.12	0.91	4.49	0.57	0.13	11.39	1.13	1.00	III
III	j	0.98	2.74	0.00	0.91	-1.21	5.84	1.12	0.43	1.22	5.90	1.13	0.44	14.49	2.26	1.79	III
III	k	-0.77	0.01	2.91	0.63	-1.51	7.38	1.85	0.92	1.48	7.24	1.76	0.86	14.63	6.51	2.41	III
	Avg.	0.03				-0.55				0.55						2.17	

Step 2 $A_I = -0.83$; $A_{II} = 0.94$; $A_{III} = 0.57$ $B_I = 0.90$; $B_{II} = -0.15$; $B_{III} = -1.06$ $C_I = -0.91$; $C_{II} = 0.16$; $C_{III} = 1.06$

Cluster	Area	A	$(A - A_I)^2$	$(A - A_{II})^2$	$(A - A_{III})^2$	B	$(B - B_I)^2$	$(B - B_{II})^2$	$(B - B_{III})^2$	C	$(C - C_I)^2$	$(C - C_{II})^2$	$(C - C_{III})^2$	Σsq-diff$_I$	Σsq-diff$_{II}$	Σsq-diff$_{III}$	To
I	a	-1.36	0.28	5.27	3.72	1.51	0.37	2.76	6.58	-1.51	0.36	2.77	6.58	1.01	10.79	16.88	I
I	b	-0.09	0.55	1.05	0.44	1.21	0.09	1.85	5.13	-1.21	0.09	1.86	5.13	0.74	4.76	10.70	I
I	c	-0.58	0.06	2.30	1.32	0.90	0.00	1.10	3.82	-0.91	0.00	1.13	3.86	0.06	4.53	9.01	I
I	d	-0.97	0.02	3.63	2.37	0.60	0.09	0.56	2.74	-0.60	0.09	0.57	2.74	0.21	4.76	7.85	I
I	e	-1.16	0.11	4.39	2.99	0.30	0.36	0.20	1.84	-0.30	0.37	0.21	1.84	0.84	4.80	6.66	I
	Avg.	-0.83				0.90				-0.91				0.57			

Table 6.5 (Continued)

Cluster	Area	A	$(A-A_I)^2$	$(A-A_{II})^2$	$(A-A_{III})^2$	B	$(B-B_I)^2$	$(B-B_{II})^2$	$(B-B_{III})^2$	C	$(C-C_I)^2$	$(C-C_{II})^2$	$(C-C_{III})^2$	Σsq-diff$_I$	Σsq-diff$_{II}$	Σsq-diff$_{III}$	To
II	f	0.79	2.63	0.02	0.05	0.00	0.82	0.02	1.11	0.00	0.82	0.02	1.11	4.27	0.07	2.27	II
II	g	1.08	3.66	0.02	0.26	-0.30	1.45	0.02	0.57	0.31	1.48	0.02	0.56	6.58	0.07	1.39	II
	Avg.	0.94				-0.15				0.16					0.07		
III	h	1.18	4.05	0.06	0.37	-0.60	2.26	0.20	0.21	0.61	2.30	0.21	0.20	8.61	0.47	0.78	III
III	i	0.89	2.97	0.00	0.10	-0.90	3.25	0.56	0.02	0.91	3.30	0.57	0.02	9.52	1.13	0.15	III
III	j	0.98	3.28	0.00	0.17	-1.21	4.47	1.12	0.02	1.22	4.52	1.13	0.03	12.27	2.26	0.22	III
III	k	-0.77	0.00	2.91	1.80	-1.51	5.83	1.85	0.21	1.48	5.69	1.76	0.18	11.52	6.51	2.18	III
	Avg.	0.57				-1.06				1.06						0.83	

Step 3 $A_I = -0.83$; $A_{II} = 1.02$; $A_{III} = 0.37$ $B_I = 0.90$; $B_{II} = -0.30$; $B_{III} = -1.21$ $C_I = -0.91$; $C_{II} = 0.31$; $C_{III} = 1.20$

Cluster	Area	A	$(A-A_I)^2$	$(A-A_{II})^2$	$(A-A_{III})^2$	B	$(B-B_I)^2$	$(B-B_{II})^2$	$(B-B_{III})^2$	C	$(C-C_I)^2$	$(C-C_{II})^2$	$(C-C_{III})^2$	Σsq-diff$_I$	Σsq-diff$_{II}$	Σsq-diff$_{III}$	To
I	a	-1.36	0.28	5.65	2.98	1.51	0.37	3.28	7.38	-1.51	0.36	3.30	7.36	1.01	12.22	17.72	I
I	b	-0.09	0.55	1.22	0.21	1.21	0.09	2.28	5.84	-1.21	0.09	2.30	5.82	0.74	5.81	11.87	I
I	c	-0.58	0.06	2.55	0.90	0.90	0.00	1.44	4.44	-0.91	0.00	1.48	4.47	0.06	5.47	9.80	I
I	d	-0.97	0.02	3.95	1.79	0.60	0.09	0.81	3.26	-0.60	0.09	0.82	3.25	0.21	5.58	8.30	I
I	e	-1.16	0.11	4.74	2.33	0.30	0.36	0.36	2.27	-0.30	0.37	0.37	2.26	0.84	5.47	6.86	I
	Avg.	-0.83				0.90				-0.91				0.57			
II	f	0.79	2.63	0.05	0.18	0.00	0.82	0.09	1.46	0.00	0.82	0.09	1.45	4.27	0.24	3.08	II
II	g	1.08	3.66	0.00	0.51	-0.30	1.45	0.00	0.82	0.31	1.48	0.00	0.80	6.58	0.00	2.13	II
II	h	1.18	4.05	0.03	0.66	-0.60	2.26	0.09	0.37	0.61	2.30	0.09	0.35	8.61	0.21	1.38	II
	Avg.	1.02				-0.30				0.31					0.15		
III	i	0.89	2.97	0.02	0.27	-0.90	3.25	0.36	0.09	0.91	3.30	0.36	0.09	9.52	0.74	0.45	III
III	j	0.98	3.28	0.00	0.38	-1.21	4.47	0.83	0.00	1.22	4.52	0.83	0.00	12.27	1.66	0.38	III
III	k	-0.77	0.00	3.19	1.29	-1.51	5.83	1.46	0.09	1.48	5.69	1.38	0.08	11.52	6.03	1.46	III
	Avg.	0.37				-1.21				1.20						0.76	

under step 1 in Table 6.5. Taking the square root of the answers gives the 'actual' distances between areas and clusters. However, there is no need to do so, provided we are consistent in the way we compare distances to clusters.

Unfortunately, the cluster allocations achieved at the end of step 1 (and shown in the final column of Table 6.5) are unlikely to be a good end solution. The reason is that the cluster centres to which areas are allocated strongly are influenced by the initial groupings of areas. We try and improve upon the result by again calculating the average attribute values for each cluster and relocating the cluster centres to those positions. Table 6.5 shows that after the first step, the cluster average attribute values reposition the clusters at $(-0.83, 0.90, -0.91)$, $(0.94, -0.15, 0.16)$ and $(0.57, -1.06, 1.06)$, respectively. Each area is reallocated to the cluster that it is now closest to, which may be different from before. For example, Table 6.5 shows that one of the previous members of cluster III becomes a member of cluster II.

The process of allocation–reallocation continues until clusters have a stable (or, sometimes in the case of larger datasets, approximately stable) membership. If you work through the process you should find that this convergence occurs at step 3. The movement of the cluster centres within the three-dimensional variable space and also the changing composition of the clusters during the classification process is illustrated by Figure 6.6. Remember that the result is dependent on the initial seed selection and should not be considered final. Instead, the process should be repeated with different selections to check for improved solutions.

Assuming that Table 6.5 does represent an optimized result then it is interesting to observe that the 'best' overall solution is not necessarily the best for any one particular cluster. In fact, cluster I has least variation within it at step 1 (this can also be deduced from the average 'Σsq-diff' value for the cluster at each step). Also, note that the clustering algorithm described here only minimizes the distance from any one area to its cluster centre. An alternative approach would be to try and maximize the distances between cluster centres to better differentiate between neighbourhood types.

Of course, different 'questions' produce different answers. The K-means procedure installed in a statistical package on one of our computers produced the following result: area a was a cluster by itself; the rest of the areas in our type I formed a second cluster; while the remaining areas (which we described as types II and III) were not grouped separately but placed together into the final, third cluster. If you have access to statistical software with a (multivariate) K-means function it may be worthwhile exploring what exactly it does with the data contained in

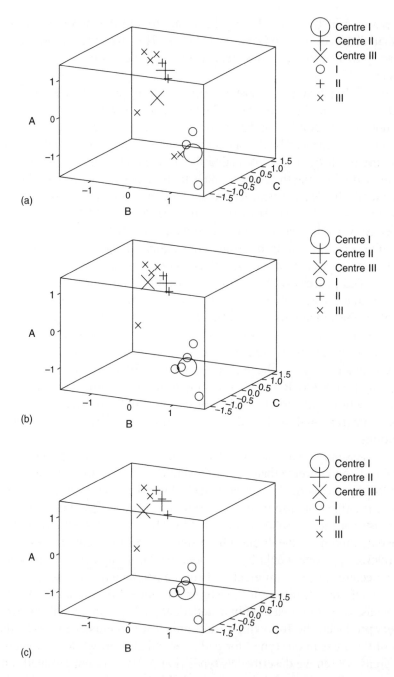

Figure 6.6 The movement of cluster centres and the changing composition of the clusters during the worked example of classification. (a) Step 1: The initial allocation of areas to seeds. (b) Step 2. (c) Step 3: the converged solution

Table 6.4. However, be aware of the limits of our example – clustering 11 areas into three clusters is not equivalent to, for example, clustering 100,000 into 10!

Before leaving our worked example, it should be emphasized that so far we have not checked to see if the variables A, B and C are duplicating underlying information. If you still have your spreadsheet open, check the (Pearson) correlations between these variables – how independent are they? If not very then try investigating the effects of assigning weights to the variables. The process of classification remains the same except for one alteration: when calculating the distance between a cluster's attribute and an area's attribute, the result needs to be multiplied by the weight given to that attribute. If attribute A is given a weight of 10, B a weight of 5 and C a weight of 2, then where before we calculated:

$$d^2 = (-1.36 - -0.68)^2 + (1.51 - 1.21)^2 + (-1.51 - -1.21)^2$$

we would instead calculate:

$$d^2 = 10 \times (-1.36 - -0.68)^2 + 5 \times (1.51 - 1.21)^2$$
$$+ 2 \times (-1.51 - -1.21)^2 \tag{6.11}$$

You might also like to explore the effects of assigning weights to areas!

6.10 Conclusion

If the process of clustering is likened to an animal then it is a very peculiar beast! It has the front legs of automation but the back legs of user intervention; eyes for data-led classification but the ears of *a priori* expectation; it feeds on a variety of data sources but generally prefers a census; displays a patchwork coat mixing the qualitative and the quantitative, the objective and the subjective; and is born as a cross-breed between art and science!

In this chapter, the process of forming a geodemographic classification has been described by a seven-stage process: selecting potential input data; bringing the data together in a database; evaluating the data; weighting the data; clustering the data; arranging the clusters into a hierarchy; and, finally, providing numerical, textual and visual summaries of the clusters. At any one of these stages the classification builder has any number of decisions or choices to make. Does this mean the process cannot be viewed as in any way objective?

One response to the question would be to quote Charlton, Openshaw and Wymer (1985, p. 74) who state:

[T]he process may seem subjective, but in reality it is no more subjective than the characteristics of other numerical methods. [I]t is objective in that the results are reproducible.

We might also cite O'Sullivan and Unwin (2003, p. 353) who suggest that things seeming a little arbitrary and not especially scientific is perhaps inevitable when applying tools that attempt to deal routinely with N-dimensional space.

However, our own 'answer' is less defensive of the subject's role within the build process and of the part the experienced developer plays in the process. Instead we take the view that the opportunities for the builder to interact with the data, try out various schemes and strategies, and to learn from past experience, successes and failures are all an important and intrinsic part of learning about the data, of 'knowledge discovery' and of looking to get a good classification of the data that is well suited to the users' needs. Whether an art or a science, geodemographic classification is an inherently creative process and is, we suggest, all the better for being so!

Summary

- Because the sources of, access to and geographic detail of geodemographic data vary internationally, classifications are usually built country by country.
- When selecting variables for clustering careful attention must be given to scale, comparability, updatability, robustness and to avoiding the duplication of existing information.
- Variable weightings can be devised using correlation analysis, minimum spanning trees, principal components analysis and prior experience.
- Two types of clustering – stepwise and K-means – are both employed by Experian during the classification process.
- Classifications are collaborated by reference to further sources of data that were not used as clustering variables.
- Cluster hierarchies are created by arranging cluster 'types' into 'groups'.
- Clusters are represented by summary labels, portraits, indicative photographs, descriptive prose, charts and maps that together

aim to explore and explain the differences between clusters. The use of labelling, in particular, has sometimes attracted critical comment.

- Clustering is never a 'one-shot' process but is objective in the sense that the developers draw on their experience, tools and available data to create a product that 'best' fits the users' requirements.

Further reading

- Everitt, B. and Dunn, G. (2001) *Applied Multivariate Data Analysis*, Hodder Arnold, London.
- Everitt, B., Landau, S., Leese, M., Ash, C. and Schober, J. (2001) *Cluster Analysis*, Hodder, London.
- Berry, M. and Linoff, G. (1997) *Data Mining Techniques for Marketing, Sales, and Customer Support*, Wiley, New York.

7
Geodemographics
Around the World

Learning Objectives

In this chapter we will:

- Look at the internationalization of geodemographics and how linkages have been established between the types of neighbourhood that are found in different countries.

- Relate this internationalization with the desire of some companies to deploy common advertising, marketing and store location techniques across all the markets in which they operate.

- Show how census, as well as other data collection and dissemination regimes, vary across the world and how these variations mould geodemographic outputs.

- Offer an overview of how geographies of neighbourhood revealed by geodemographic classification compare and contrast across the world, suggesting structural, social and cultural differences between places of which marketers needs to be aware.

- Present a case study by Peter Furness offering a comparison of selected censuses from across the world.

Geodemographics, GIS and Neighbourhood Targeting   Richard Harris, Peter Sleight and Richard Webber
© 2005 John Wiley & Sons, Ltd   ISBNs: 0-470-86413-3 (HB); 0-470-86414-1 (PB)

Introduction

In Chapters 2 and 3 we discussed the evolution of geodemographic classification from public policy research into a commercial industry. Geodemographic classifications were first developed for public policy applications during the late 1970s but were quickly launched during 1978/9 as tools for target marketing in the USA and the UK. The US system was developed by Claritas and branded PRIZM. The UK equivalent, which was adapted by CACI from an earlier system, came to be branded as ACORN. Since 1979 the methodological approaches used by PRIZM and ACORN have been the model for developing further neighbourhood classifications, not just in the USA and UK but for an ever increasing number of counties across the world.

Indeed, by the turn of the millennium, Weiss (2000, p. 33) was able to write:

> Throughout the world, there's remarkable similarity in the way businesses are using the cluster technology – for analyzing trading areas, profiling customers, and driving media strategies. The increasing globalization of culture is also prompting multinational companies to look at clusters as a common marketing language to reach customers across many borders.

Using this language of marketing, Weiss portrays not only American lifestyles in terms of this 'clustered world' but also makes comparisons with Canada and many other parts of the world in a search for the international clusters that characterize the 'global village'.

Here our aims are more modest. We begin by looking at the internationalization of geodemographics, noting some of the many countries where classifications have now been produced, some differences in the data sources and geographical frameworks upon which they are built, and paying particular attention to some of the similarities and differences in census taking across the world and how these impact on the geodemographic information provided. Having done this we look at some visualization and multimedia tools from classifications produced around the world to offer an overview of a few of the common and differing geographies of neighbourhood that are found worldwide. These especially are relevant to companies that seek to standardize their use of geodemographics internationally, to develop consistent transnational marketing campaigns for example.

7.1 The internationalization of geodemographics

Across Europe, neighbourhood classifications have been built by Claritas in France, Germany, Netherlands, Spain and the UK, and by Experian in Belgium, Denmark, Finland, Germany, Greece, Ireland, Italy, Netherlands, Norway, Spain, Sweden and the UK, either to support wholly owned subsidiaries or as the basis for joint-venture or franchise operations. In France, a further neighbourhood classification was built by Coref (a company which was later acquired by Experian) and in Italy by Sarin, a division of the Italian Yellow Pages (a telephone directory service).

Neighbourhood classifications have also been built and promoted in Canada by Compusearch, a company recently acquired by MapInfo – a GIS vendor. Peruvian and Brazilian systems have been developed with the support of Experian. South African marketers have for many years had access to a classification trading under a number of different brands, including Mosaic. In the US, classifications include Mosaic, PRIZM (Claritas) and PSYTE (Polk).

Neighbourhood classifications are used extensively by marketers in Australia and New Zealand. Systems for Hong Kong and Japan have been developed by Experian, initially to support the international expansion of Marks and Spencer (a British-based retailer). More recently Smartal Systems have built a system for Hong Kong and launched a system covering the three most prosperous city regions in China – Shanghai, Beijing and Guangzhou.

The principal motivation for developing these systems has been to support commercial rather than public-sector research applications. In most of these markets the business model broadly is similar as also is the manner in which the classifications are built, documented, visualized and positioned. Although, the results of these classifications are largely unpublished the visualization tools and multimedia guides that accompany geodemographic information systems can reveal interesting information about the social structure and neighbourhood composition of different countries, providing a useful basis for comparison.

Table 7.1 lists markets according to the availability of such visualization tools and, where they are available, gives an example of a national market research survey operation which supports the tabulation of consumer behaviour data using a cluster-based approach. Although these systems extensively are marketed in their 'home' countries, there is an increasing

Table 7.1 Examples of geodemographic tools from around the world

Country	Availability of visualization material	Example of geodemographically coded market research survey	Contents
Australia	Available	Roy Morgan	Attitude and opinions, FMCG, sporting interests, demographics, household appliances, personal products, automotive, holidays, media
China	Not available	Target Group Index	Products, media, brands
Denmark	Not available	Gallup	Interests, attitudes, consumption by category/product/band, media, travel and transportation, business to business, retail
Finland	Available	Taloustkimus	Ownership of consumer durables and media consumption
France	Planned 2004	Secodip	Drugs/fragrances/hygiene and textile – expenditure per household for consumer goods
Germany	Available	GFK	FMCG
Greece	Not available	Target Group Index	Demographics, products ownership/intention to buy, lifestyle information, media, travel, car ownership, financial services
Hong Kong	Not available	Nielsen	Consumer durables, FMCG
Japan	Not available	None	None
Netherlands	Available	NIPO	Media
New Zealand	Not available	Roy Morgan	Media, attitude and opinions, brand preferences, shopping habits, activities and interests
Northern Ireland	Planned 2004	Target Group Index	Media, products, brands
Norway	Not available	Gallup	Consumer products, media, channels
Republic of Ireland	Planned 2004	Ireland Life Survey	Financial services, telecomms, technology, media, lifestyle, activities and hobbies
Spain	Available	Estudio General de Medios	Media, shopping, finance, cars
Sweden	Available	Orvesto Konsument	Consumer behaviour and preferences – interests, possession of durables, media habits, spending
UK	Available	Target Group Index	Products, media, brands
USA	Available	Simmons	Services, products, media usage

demand from large global businesses to deploy 'overseas' systems within their international headquarters or regional marketing offices. For example, there is a team at Marks and Spencer's international headquarters, in London, which applies overseas geodemographic systems for analysing its non-UK markets.

Comparisons across national boundaries are also made possible by the development of products such as Mosaic Global, an ambitious undertaking that covers nearly 800 million of the world's most affluent consumers, across 17 countries, identifing 14 distinct types of residential neighbourhood, each with its own set of values, motivations and consumer preferences (www.business-strategies.co.uk). Such products (which also include EuroDirect's CAMEO International classification: www.eurodirect.co.uk) reflect the desire of companies such as McDonald's, Ford and Citibank to deploy common advertising, marketing and store location techniques across all the markets in which they operate and therefore to look to a single supplier of geodemographic classification just as they would to a worldwide supplier of merchandise, parts, advertising, accountancy or other business services. In recent years companies including Marks and Spencer, McDonald's, Ford and Ikea have taken the decision to standardize, at an international level, the manner in which they use geodemographic systems for applications such as retail site location and strategic market segmentation. This is consistent with the undertakings of major international brands to appoint global advertisers to develop consistent transnational marketing campaigns.

Case study: a brief comparison of selected censuses from across the world

Peter Furness, Peter Furness Limited
 The text is a shortened version of a chapter published in Dugmore and Moy (2004) and is here reproduced by permission of The Stationery Office Ltd.

From its foundation the United Nations (UN) has recommended national governments hold population censuses at the end or beginning of each decade. Most countries thus undertake population and housing censuses at least once every 10 years. The UN Statistical Commission estimates that 165 countries have conducted a census in the course of the last decade. These censuses have covered approximately 95% of the world's population.

The methods used for census taking cover a wide spectrum, ranging from the traditional questionnaire-based census through to those carried out using administrative registers. In between are combinations of the two, supplemented in some cases by various types of sample survey, either in conjunction with the census (sample enumeration) or as separate activities (household surveys).

The traditional census – typified by the UK and many other countries including major economies such as the USA, China and Japan – involves the distribution of census questionnaires which are then collected by enumerators or by a combination of enumerators and 'post back'. Some countries such as the USA, Switzerland and Spain are experimenting with the internet as a data collection medium.

The traditional census has three main advantages. First, the national statistical organization has control over the operation which means that it can be managed to maximize the statistical potential of the data collected. Secondly, the traditional census tends to have wide topic coverage. Third the public awareness generated for official statistics in general engenders a sense of national participation and highlights the importance of objective and timely statistics for society at large.

The main disadvantages of the traditional census are costs, the infrequency of enumeration and associated organizational headaches, and the problems of accurately enumerating certain subgroups of the population such as the young mobile, people living in 'ghettos' or 'shanty towns' and some other marginalized sections of the community.

For the cost reason, many countries carry out a sample enumeration in conjunction with the census to collect more detailed information on a separate, longer questionnaire. This is often referred to as the 'long form', as in the USA, Canada and China, and is distinct from the 'short form' to be completed by all residents and used to compile baseline demographics including number of persons, households, gender, age and so forth.

Separate from decennial sample enumeration (the long form), some countries employ sample household surveys as an integral part of their census operations. Household surveys are the most flexible of the census data sources. In principle, almost any subject can be investigated through this method. France plans to replace its traditional census with a continuous, rolling survey that will cover the whole country over a period of time rather than on one particular day. This will be used in conjunction with small-area modelling techniques to provide annual counts on a continuous basis. Germany conducts a rolling 'microcensus' of 1% of households annually, organized on a partial rotation basis so that every sampled household remains in the sample for four years. In the UK the traditional census is under review:

> The Census provides vital information for Government and business planning that is widely used in the public and private sectors, but robust evidence to justify the expenditure involved in undertaking the Census was not

available [to a Government committee]. We recommend that any future Census should be justified in cost-benefit terms. Moreover, before such a decision is taken, an evaluation should be undertaken of all alternatives to a full Census, such as relying on administrative records, conducting a rolling Census, reverting to a simple headcount, or doing without a Census altogether (Treasury Committee, 2002).

The preceding quotation makes reference to the increasingly used data contained in administrative registers. Many social statistics are produced as a byproduct of administrative processes. Data from different registers can be merged (provided they have a common georeference such as postal address) thus reducing the burden on the respondent by making use of whatever data is already in the system. Other benefits include the avoidance of costly 'in the field' data collection and a greater degree of control over the timing and frequency of statistical reporting.

However, there are also disadvantages of using administrative registers. The content and detail may substantially be less than that available from questionnaire-based surveys and there may be data quality and error propagation problems, which make the merging of registers problematic. There is also the significant challenge of convincing the public about the confidentiality of the merging and reporting processes – issues of data privacy and protection; or, 'dataveillance' as Monmonier (2002) describes it.

Nevertheless, many developed countries, particularly in Europe, have been developing such a capacity for a number of years. Nordic countries – Denmark, Finland, Iceland, Norway and Sweden – have relied on register-based statistics for their data needs since the 1960s and have used demographic information from central population registers since the 1970 censuses. Other countries in Europe, like Austria, Belgium, Luxembourg and Switzerland are in the process of transition from traditional to register-based censuses. Countries such as France, Germany and the Netherlands are developing a combination of various types of register and sample surveys to produce census-type data.

Table 7.2 provides an overview of the current census methodologies adopted by the world's 10 largest economies. For further information about censuses from across the world refer to Office for National Statistics (2003) – which is available from www.statistics.gov.uk. In less wealthy nations where the cost of providing a census or other types of data infrastructure is prohibitively expensive, the use of remotely sensed imagery has been suggested (Baudot, 2001).

As for the future, it is clear that the nature of the census and the manner in which it is conducted is changing in many countries. There are several factors involved. First is the impact of technology on the traditional

Table 7.2 Census summary for the world's top 10 economies

Country	GDP (2002 in 10^{12})	Population size (2001 in millions)	Most recent population census	Methodology	Website link
USA	10.40	285.9	April 2000	Traditional with future methodology under review	http://www.census.gov/
China	5.70	1285.0	November 2000	Traditional	http://www.stats.gov.cn/english/index.htm
Japan	3.55	127.3	October 2000	Traditional	http://www.stat.go.jp/english/index.htm
India	2.66	1025.1	March 2001	Traditional	http://www.censusindia.net/
Germany	2.18	82.0	N/A	Register based plus sample surveys	http://www.statistik-bund.de/e_home.htm
France	1.54	59.5	March 1999	Traditional moving to register based plus sample surveys	http://www.insee.fr/
UK	1.52	59.5	April 2001	Traditional with methodology under review	http://www.statistics.gov.uk/
Italy	1.44	57.5	October 2001	Traditional	http://www.istat.it/English/index.htm
Russia	1.35	144.7	October 2002	Traditional	http://www.gks.ru/eng/
Brazil	1.34	172.6	August 2000	Traditional	http://www.ibge.gov.br/english/default.php

Sources: GDP data from www.countryrankings.com; population sizes from the United Nations (Infonation).

census. This is being felt in all aspects of the census process: from planning and management; through data capture, processing and storage; to data dissemination. It is fuelled by tools such as GIS, intelligent character recognition (to make automatic the process of data collation), the internet as a data collection and dissemination medium, and by new analysis and reporting software.

Second is the improving range of data sources for a register-based census. As we have seen, an increasing number of countries are adopting the register-based approach to supplement or even replace the traditional census. Pressures to implement population registers where these are not currently available (for example, from increasing concerns about the need to tighten control over citizenship and entitlement to services) together with the availability of new and highly innovative data sources (such as the Cooltown Project from Hewlett-Packard, www.cooltown.com), will tend to fuel this trend.

Thirdly, in a world with accelerating population and socio-economic change, the traditional decennial census is becoming less and less relevant after the first few years following a census. With cost the major barrier to a more frequent traditional census, there will be greater pressure for the use of register-based data and frequent sample surveys to supplement or supersede the census.

Against this, the general public is becoming increasingly concerned about data privacy and the restrictions on civil liberties that population registers might engender (see Chapter 9). This may well impact on the scope of administrative data sources that can be mobilized for census-related activities and the speed with which change can be effected.

7.2 Census data sources – some differences in what is asked and where

Of the 21 countries and one regional city (Hong Kong) for which geodemographic classifications have been built, census statistics have been used in all but the Netherlands and Germany. In all of the remaining countries, other than France and Canada, geodemographic classifications have been updated on a 10-year cycle according to the frequency with which census statistics are released.

The census questions that are asked in the different national censuses broadly are similar in scope, covering: age, ethnicity and household characteristics; housing; and various measures of affluence and deprivation including education, occupation, industry and car ownership. Despite these broad similarities there are interesting and important differences in each market, often reflecting differences in the level of economic development. These differences often have a significant effect on the shape of the eventual classification.

In Australia and the USA, for instance, the census provides unusually detailed information on rents, house values and personal and household incomes, making it correspondingly easier to differentiate neighbourhoods characterized by high incomes and by high housing costs – which predominantly occur in the larger cities – from less well-off neighbourhoods in smaller towns and rural locations. The Australian census is also unusual in that it covers access to the internet, a very useful proxy for the 'Urban Intelligence' group of neighbourhoods (defined by Mosaic) and a highly relevant measure of social inclusion. The Australian census is also unique in that it asks the general subject in which degrees were taken. From this information it has been possible to identify that the 'Café Society' neighbourhoods of inner Sydney and Melbourne have disproportionate numbers of people who have graduated in arts rather than scientific subjects.

The Brazilian, Peruvian and South African censuses are, by contrast, much more focused on topics relating to access to utilities such as water and electricity that differentiate shanty towns from mainstream developments. These censuses cover whether or not households have access to running water, electricity or mains sewerage. In these countries, as in China, the censuses also have quite detailed information on literacy. The Brazilian census has particularly detailed information on the consanguineous relationships of the often extended households that live together. Like nineteenth-century British censuses, it provides information on the geographical distribution of domestic servants (the sort of information that Charles Booth used to construct the first census-based classes of social economic groups within the UK: see Chapter 2, Section 2.1).

A distinctive feature of the Japanese census is the coverage of the amount of time it takes to get to work, whereas the British census provides detail on the mode of travel of work. Information on these topics improves the ability of the neighbourhood classification to differentiate neighbourhoods in metropolitan and regional centres from those in smaller towns.

The Chinese census features a number of distinctive topics. One important table in the Chinese census is a resident's 'hukou'. In China everyone is registered with a local community, entitling them to certain rights. The distribution of the local population according to whether their registration (hukou) is local and, if not, how long they have been resident at their address is a useful indicator of the presence in a neighbourhood of students or of temporary workers from other regions of China. Another distinctive feature of the Chinese census is whether the dwellings that people live in are wholly or only partly residential. This makes it quite easy to identify areas of shop-houses in the centres of busy commercial districts. In China the proportion of households using wood as fuel for heating and cooking can be seen from the statistical profiles to be a very effective indicator of traditional rural areas.

Scandinavian and Spanish censuses provide detailed information on the local concentration of second homes while the British census provides unique insights into whether homes are terraced, semi-detached or detached. The British census has always provided detailed information on housing disadvantage, including the proportions of households sharing access to a bath or an inside toilet, or living with more than one person per room. The UK census does not require its respondents to know the date of construction of their homes, unlike the Chinese and Hong Kong censuses. The Chinese census is unique in requiring people to specify the number of square metres of their accommodation as well as the number of rooms.

In addition to differences in the information sought, different censuses have different scales of data output. For instance, Finland, New Zealand, Norway and the UK release data for relatively small geographical units; Spain and Sweden for relatively large ones.

7.3 Differences in the availability of non-census data sources

In many countries geodemographic systems exclusively are built using small-area census statistics. However, where these are not available, as for instance in Germany and the Netherlands whose citizens, since 1945, have resisted proposals for the collection and dissemination of demographic data at small-area level, neighbourhood classifications have been constructed using an eclectic array of data from a variety of sources. Particularly important in Germany has been a field evaluation of individual buildings and/or streets, which provides information on their age, physical form and building condition. Telephone directories and car registration files also provide important inputs to the German system.

In the Netherlands, information used to build neighbourhood classification originates in part from the sales records of Wehkamp – Holland's leading mail order company – and from the accumulation of many years of information from conventional market research interviews. In both Germany and Holland the availability of data for very detailed geographic units counterbalances the relatively narrow range of topics otherwise included in the databases used to build the system.

In Spain, the high level of granularity (low resolution) of the census output statistics makes it necessary to supplement them with a small set of key variables obtained from the electoral register. This is not the case in France, Italy, Ireland and Greece where census statistics are accessible at a much more detailed geographic level and therefore do not need to be supplemented by non-census data sources. An electoral register is less critical for the success of New Zealand and Australian classifications but has been used to increase the geographic level to a finer level. Indeed the availability of cadastral data for individual building units in New Zealand allows the classification to operate right down to the building level.

7.4 Variations in the detail of the postal delivery system

A third critical consideration when commercializing classifications in different markets is the granularity of the postal system. Where postal units (such as unit postcodes or ZIP+4) are very fine, as for instance in the USA, the UK and the Netherlands, it is very easy for users to identify the geodemographic code of customers or a client by using a look-up file linking postcodes to classifications. However, in countries where postcodes do not exist at all (such as Hong Kong) or where they can cover very large areas (such as in Germany or Spain), it is necessary to develop or apply address-matching software to customer or survey records in order to append geodemographic codes. This step clearly adds a level of complexity and cost to the analysis and targeting process, which in the past has been a significant barrier to the adoption of neighbourhood classifications in these countries.

7.5 Geographies of neighbourhood worldwide

As we have seen, between countries in which geodemographic classifications have been built there is variation in the sources of local statistics that are available, differences in the level of geographic detail at which these data are released, differences in the granularity of the mail delivery systems used to identify geodemographic codings and, perhaps most critically, variations in the nature, emphasis and dissemination of census information. Furthermore, important social, cultural and political differences between nation mean that we would not expect to see any two countries having exactly matching types of neighbourhoods. Population size also makes a difference: in small and relatively homogeneous countries, such as Ireland, Finland and Hong Kong, we might expect to find fewer distinguishable types of neighbourhood than in very large and populous countries such as the USA, Brazil and China that cover different climatic zones and which, in the case of the USA and Brazil, contain substantial minority populations with quite distinctive behavioural characteristics.

Despite these differences and variations, marketers have gone some way to making connections across the various neighbourhood types, which then permit international comparisons to be made. For example, Weiss (2000) suggests that (at the time he was writing) the 'Blue Blood Estates' of the US PRIZM classification, the 'Suburban Executives' of the Canadian PSYTE classification and the 'Clever Capitalists' of the GB Mosaic classification all fell, with some other neighbourhood types, under the broader socio-economic banner of 'Elite Suburbs'. Similarly, US 'Urban Gold Coast', the 'Canadian Establishment' and UK 'Studio Singles' might all be considered as 'Upscale Urban Areas'.

The explanations Weiss gives for such connections are associated with general notions of globalization and the export of American culture overseas. Yet, there arises somewhat of a contradiction in that he is also concerned to illuminate the fragmentation – not homogenization – of American and other national lifestyles. The suggestion appears to be that cultures are both homogenizing but also fragmenting in broadly similar ways across the more affluent countries of the world, perhaps reflecting the rise of consumer societies offering similar types of branded goods and services across the globe, and perhaps reflecting notions of self-similarity, of patterns repeating at various scales of analysis – a fractal perspective that has been employed in analyses and writings of urban settlements (Batty and Longley, 1994; Soja, 2000).

There are limits to the sorts of connections that can be made between the neighbourhood classifications of differing nations and these need to be understood and recognized. Most particularly (and notwithstanding variations in data sources) the broad method of classification is the same in each country. This offers standardization and comparability but also tends to represent the world in terms of what matters to business and commerce, not what may actually be of importance to the people of those countries, their particular social contexts, values and aspirations. In other words, attempting to portray 17 countries in terms of only 14 neighbourhood types may reasonably lead to accusations of oversimplification, of lacking cultural sensitivity and specificity.

On the other hand, important differences – in consumer behaviour at least – can be revealed by the common analytical base provided by geodemographics. For example, Weiss (2000) cites the example of Bang and Olufsen who found that consumer behaviours differ in Britain from countries like France and Spain. Specifically, the British were prepared to travel further distances to a Bang and Olufsen store, meaning the catchment models based on a British market did not export well. Often it is these 'exceptions that prove the rule' that are the most interesting. A virtue of the geodemographic approach is that classifications can at times highlight cluster types that simply do not match with ones for other countries – that are genuinely unique. An example of this is the 'Shack and Shanty' type of South Africa, which comprises about half of the country. Its name and prevalence is indicative of the sharp economic disparities that exist in that nation.

It is quickly evident from even cursory inspection of the visualization material supporting geodemographic products that there are a number of interesting similarities and differences in the patterns and classes of neighbourhood type revealed by the classifications produced for different countries. These in part reflect the methods of classification yet also shed light on some underlying structural, social and cultural differences between places. This perspective is consistent with our long definition of geodemographics (Chapter 1): the analysis of socio-economic and behavioural data about people, to investigate the geographical patterns that structure and are structured by the forms and functions of settlements. To quote what we said in a previous context of measuring deprivation (Chapter 2, Section 2.3), what geodemographic profiling comes to represent exists at the interplay between the real world, and the set of methods and meanings used to make sense of it.

The similarities and differences in neighbourhood structures, which are revealed by geodemographic comparisons, are highly relevant to large international agencies. They suggest how and where otherwise uniform global branding campaigns need to be adapted to take into account particular features of the demographics of individual markets. With that in mind, we proceed to describe some of those differences, mainly limiting our discussion to those between northern and Mediterranean Europe, between Europe and the Americas, and between developing and advanced industrial societies. We are aware that by some academic theorizations our descriptions are somewhat partial and oblique but still, we hope, impart a flavour of how neighbourhood geographies compare across national borders and are conceptualized in some aspects of marketing.

7.5.1 Comparing neighbourhood groups of northern and Mediterranean Europe

When the geodemographic profiles of northern European markets are compared with those of Mediterranean Europe, we find that in the former there is a much higher proportion of clusters characterized by low-rise houses as opposed to high-rise flats. Whereas in northern European classifications almost all high-status neighbourhood categories consist of low-rise houses (Mosaic UK's 'Global Connections' type is an exception), higher proportions of the 'better-off' clusters and population are to be found in inner city locations in Spain ('Classic Bourgeois' neighbourhoods), Italy ('Blue Ties') and Greece. The 'flight to the suburbs' and the decentralization of cities that has, until recently at least, generally characterized population movements within the UK has not happened to the same extent elsewhere (the UK's Urban Task Force report frequently cites the higher population densities of Mediterranean cities as the model for sustainable urban living: Urban Task Force, 1999). These differences impact on leisure pursuits as measured by lifestyle surveys – the popularity of gardening in the UK, for example.

A second contrast emerging from northern European and Mediterranean classifications is the scale at which spatial differences in levels of prosperity occur. While in southern Europe the major differences in levels of income and prosperity occur between clusters concentrated in neighbourhoods in metropolitan and regional centres on the one hand, and in clusters concentrated in neighbourhoods in small towns and rural areas on the other, in northern Europe and particularly in Britain, the clusters are much less regionalized. In these systems, high- and low-income clusters tend to appear much more equally in cities of different size and there are relatively fewer urban areas which contain exclusively high- or exclusively low-income neighbourhoods.

Marketers, such as those in American Express or Vodafone, when profiling the adopters of new products or of new channels of communication, consistently have found early adopters in southern Europe to be much more strongly associated with metropolitan neighbourhood types.

By contrast in northern European markets it would appear that a much higher proportion of early adopters live in suburban and semi-urban locations and around second tier and provincial cities, as well as major metropolitan centres. The perception of many brand marketers is that the spatial obstacles to the diffusion of innovatory propositions is much weaker in northern than in southern Europe where rural provinces are perceived as being more traditional in their orientation. The contrast in the textual vignette describing Italian 'Soil and Sun' and the UK's 'Greenbelt Guardians' demonstrate very clearly differences in terms of education, incomes, use of financial services and of different media channels.

A third contrast between northern and southern Europe is the influence of the family. Northern European classifications demonstrate the concentration within provincial cities of large areas populated by young single people who either live on their own or in shared accommodation. Many of these types of neighbourhood (such as the Dutch 'Young Singles in Inner City Surroundings') are located close to universities, to which children go to, away from home. However, larger cities – London in particular – have many areas dominated by childless households, whether singles, sharers, co-habitees or recently married. By contrast, in Spain, Italy and Greece (and elsewhere in southern Europe), few people other than the military live in neighbourhoods characterized by young singles. These are not cultures that spawn neighbourhoods (such as 'Counter Cultural Mix') of young singles sharing large, old Victorian houses subdivided into bedsits or small flats of the sort featured in the film *Notting Hill*. Such distinctions are useful to marketers for suggesting the most appropriate manner of promoting products to young adults.

7.5.2 Comparing neighbourhood groups of Europe and the Americas

The neighbourhood patterns revealed by classifications of the USA and Canada generally are closer to those of northern Europe than to those of southern Europe. In the USA and Canada, as in Scandinavia and Germany, geodemographic profiling suggests that the most affluent neighbourhoods tend to be close to the outer ring of most of the largest cities – an enduring realization of the Burgess model of 1920s' Chicago. Conversely, inner urban areas, other than in New York and Boston, have had difficulty retaining more affluent households, except in neighbourhoods dominated by households without children (such as US Mosaic's

'Upper Bohemia'). However, unlike northern Europe and perhaps due to the huge territorial size of the USA relative to European countries, there is a greater tendency for the US clusters which contain the lowest income neighbourhoods to be located in remote rural areas, particularly the south (Mississippi and Arkansas, for example). As in southern Europe, there is a tendency for the more affluent, family-oriented neighbourhoods particularly to be concentrated in specific regions which are closely networked into the global economy and whose high levels of education and innovation have resulted in particularly high land and housing costs and a high reliance on knowledge industries (such as US 'Latino Norte'). In the US classifications it is common for more than one neighbourhood cluster to be associated almost exclusively with the state of California.

Comparing levels of income, unemployment and ethnicity in the different clusters, it is evident that US neighbourhoods are significantly more polarized than European ones in terms of both demographic and consumption patterns. The corollary of this greater variability between types is a lower level of diversity within types and, indeed, within statistical areas. Geodemographic clusters are much more ethnically segregated in the USA and there are a number where particular ('non-white') ethnic groups make up a majority of the population. This contrasts with the situation in Europe where no clusters are dominated by any particular 'minority group'. Accordingly, US classifications support more ethnically determined categories than European ones, with ethnic clusters that are specific to rural as well as urban areas.

7.5.3 Neighbourhood geographies of Latin America

In both Brazil and Peru it is evident from maps of the various neighbourhood clusters that more affluent neighbourhoods remain particularly focused on well-established near-central locations, often with an attractive historic environment. Most peripheral locations are occupied by recent, poor migrants from rural areas. Neither Brazil nor Peru have areas of young singles or sharers: neither social mores nor the housing stock is conducive to the emergence of these areas and most students continue to live at home with their parents and to study at local universities. Problems with security and the absence of public housing mean that the more affluent, as a rule, favour neighbourhoods of apartments over neighbourhoods with villas. These neighbourhoods, such as Miraflores in Lima, tend geographically to be more concentrated than is the case in Europe and they tend to be located along a limited number of radial corridors. These neighbourhoods have, over time, attracted the offices of international companies with the result that many people from poor neighbourhoods commute from inner city neighbourhoods along these radial corridors to gain employment in the more wealthy suburbs.

Another pattern that is distinctive to Latin America is the tendency for new neighbourhoods to improve their status over time. Whereas in the USA there is a tendency for the highest income groups to live in the newest neighbourhoods and for neighbourhoods to lose status the older they get, in many Latin American 'favellas' it is common for the newest rural migrants to live in the newest neighbourhoods and for these poor neighbourhoods to improve their status as their new inhabitants succeed in establishing themselves in the urban economy. The Peruvian geodemographic classification shows very clearly how the parts of cities newly built by Andean migrants first become legally incorporated, then acquire water and drainage, then are cabled for electricity. Roads become paved. Shacks are reconstructed in more durable materials and their owners, rather than move out and 'up' as would ethnic minorities in the larger Texan or Californian cities, instead build a second upper storey with their emerging wealth. Climate and soils likewise militate against the exodus of the rich into commutable villages, as would be case in the UK, Netherlands, Belgium and France. The leisure focus of the wealthy is more on the country club than the country retreat. Neighbourhoods of second or retirement homes appear wholly absent.

7.5.4 Neighbourhood geographies of 'the east' and 'the west'

Australian classifications provide evidence of the similarity between Australian neighbourhoods and those of the USA, the UK and northern Europe with a high number of clusters characterized by footloose singles and a much higher concentration of wealthy, globally connected and well-educated (though not necessarily high income) residents living in inner city neighbourhoods, particularly where there is outstanding architectural or environmental heritage (such as Australia's 'Champagne and Chardonnay' neighbourhoods). Australia and New Zealand, like Britain, have retirement rather than vacation seaside clusters (such as New Zealand's 'Sun Seeking Seniors') but, like the USA and Scandinavia, have no high-status rural commuter neighbourhoods. Australia's neighbourhoods of public housing, like Britain's, have high concentrations of poor, 'non-white' ethnic groups.

In both China and Japan the neighbourhood classifications clearly differentiate self-reliant populations who live in neighbourhoods (such as Japan's 'Working Village Families') characterized by old people and low incomes yet at the same time living in spacious, mostly owner-occupied accommodation (where there is very little unemployment), from neighbourhoods in large cities that tend to contain younger populations that are more likely to live in flats which they rent and who travel long distances to work in service industries. In both China and Japan there are clear differences between the mix of neighbourhoods in Beijing, Shanghai and Tokyo, on the one hand, and of second-tier cities where the population is less well educated and much less likely to work in service industries. Late marriage rates, low fertility and delayed childbirth are a distinctive feature of the affluent neighbourhoods of the top-tier cities as they are in global cities of Europe and the USA.

Despite national – even continental differences – a number of common dimensions are apparent in all classifications. Even in Sweden, where disparities in income after tax are relatively low, and in China, where housing historically has not been allocated on the basis of market forces, there is still a tendency for people of different levels of education to find themselves as neighbours of others of similar status. This tendency to 'flock together' remains apparent even in areas of housing which originally were developed for public renting – whether in the UK, Hong Kong or China, these tend still to be differentiated on the basis of status.

In every market there are some clusters that are highly connected to global networks, whether financial, academic or cultural, and which attract people who are the most interested in new ideas, in international trends and for whom diversity and complexity is an attraction rather than a repellent. These neighbourhoods (such as UK 'Counter Cultural Mix' and 'New Urban Colonists', or Japan's 'City Flat Dwellers') particularly are associated with political capitals, banking centres and prestige university cities. Residents in these neighbourhoods have a high thirst for media and have more cosmopolitan worldviews. People who live in such areas will often find themselves more at home in similar neighbourhoods in other countries than they would in other sorts of neighbourhood in their home country.

By contrast, the types of neighbourhood which are the most quintessential of a country, where tastes and traditions are most old fashioned and traditional, are likely to be found not just in the countryside (such as Norway's 'Mountain Farmers') but also in clusters characterized by older, often terraced and often privately rented housing for workers in old established mining and heavy industrial areas. Such areas are particularly conservative in their tastes for food, in their sporting preferences, in their disregard of international brands and, sometimes, in their dislike of 'incomers' from other types of neighbourhood. Family and the community provide an important source of support in these areas, where people marry and have children when they are particularly young. The South Wales valleys (such as UK 'Industrial Grit'), Appalachia, the Belgian coalfield (such as Belgian 'Industrial Tradition') and the Basque region provide high concentrations of neighbourhoods of this sort.

In most countries there is an apparent divide between the neighbourhoods in which the more cosmopolitan 'Chattering Classes' live and those, often with a more scientific education, who work for large international organizations with scientific and technical orientations. More often located on the outer rim of major growth poles, these residents have a much more consumerist lifestyle than their counterparts in inner cities and are more likely to live in new housing, to be adept at the use of new technologies and to commute by car, to work in high tech businesses located in office parks. Examples are UK 'High Technologists', Italy's 'Silicon Heartland', France's 'Rising Materialists', Finland's 'Silicon Valley' and China's 'Upcoming Elites', Helsinki (Finland), Stockholm (Sweden), Reading (UK), Utrecht (The Netherlands) and Seattle (USA) all have high concentrations of residents in this type of neighbourhood.

7.6 Conclusion

In this chapter we have shown that between the many countries in which geodemographic products are available there are differences in the sources, specification and geographic detail of the data used to build the classifications. Despite these differences, connections have been made between apparently similar types of neighbourhood in different countries. Making such links can be useful to identify commonalities and differences in neighbourhood patterns which, in turn, help to reveal underling structural, social and cultural trends. They also help to identify neighbourhoods that simply cannot be matched and, in this way, are unique to particular nations.

Such comparisons and the conclusions drawn from them are partial and somewhat superficial. However, for marketers there are benefits to be gained by interpolating results from neighbourhood types in one market to near-matching neighbourhood types in other markets. This interpolation is likely to be more reliable at the product category level (i.e. for purchase and ownership of cars), than at the brand level (i.e. the purchase of a particular brand of car, such as Ford or Honda). It is always important to remember the cultural differences that will exist between residents living in notionally similar areas but in different countries. And to remember that especially (but not exclusively) in very large countries such as the USA, China and Brazil, behaviours associated with a given type of neighbourhood may display considerable regional variety. A good example is the consumption of food products which themselves are a reflection of differences in climate, soils and produce that historically could be easily grown or raised. However, other (perhaps less obvious) attitudes, preferences and behaviours are also often regionalized.

What this all adds up to is a sense of perspective and a realization that geodemographic methods are one tool among many ways of constructing geographical knowledge. They make no pretence to explain fully the world, only to help us understand it a little better, in a way that is useful for private- and public-sector decision making. It will be useful to keep this 'sense of place' in mind when we come to critique geodemographics in the following chapter.

Summary

- Geodemographic methods increasingly are used throughout the world for analysis, profiling and to inform marketing.
- Neighbourhood classifications have, for example, been built in France, Germany, the Netherlands, Spain, the UK, Belgium, Denmark, Finland, Greece, Ireland, Italy, Norway, Sweden, Canada, the USA, Australia, New Zealand, China, Japan, Brazil and Peru.
- Methods and sources of data collection vary across the world.
- Most countries undertake a census every five or 10 years. Methods of census include distributing the same questionnaire to all households, using survey techniques, fusing datasets and maintaining rolling registers.
- Some similarities and differences in neighbourhood structures can be revealed by comparing geodemographic outputs (profiles, visualizations and multimedia tools) from across the world. Such comparisons are useful to large international businesses and agencies that wish to deploy common advertising, marketing and store location techniques across all the markets in which they operate.
- Mosaic Global is an example of an international geodemographic product, classifying 800 million consumers, across 17 countries into 14 distinct types of residential neighbourhood.

Further Reading

- Allen, D. (1968) *British Tastes: an Enquiry into the Likes and Dislikes of the Regional Consumer*, Hutchinson, London.
- Weiss, M. (1994) *Latitudes and Attitudes: an Atlas of American Tastes, Trends, Politics, and Passions, from Abilene, Texas to Zanesville, Ohio*, Little, Brown, New York.
- Weiss, M. (2000) *The Clustered World*, Little, Brown, New York.

8
'But Does It Work?'
Geodemographics
in the Dock

Learning Objectives

In this chapter we will:

- Consider the representations of people and places given by geodemographic systems.

- Further consider how the twin issues of the ecological fallacy and the modifiable areal unit problem affect the analysis of geographic information.

- Question whether apparent geodemographic differences between neighbourhoods are as much a consequence of the analytical methods used to find them as they are of any real-world patterning.

- Present case studies showing how, where and why geodemographics has been used with success.

- Emphasize the relative simplicity, ease of understanding and portability of geodemographics, and its role as a tool for organizing and comparing geographic datasets.

- Present a case study used to validate the geodemographic approach, authored by Barry Leventhal of Teradata, a division of NCR Corporation.

Geodemographics, GIS and Neighbourhood Targeting   Richard Harris, Peter Sleight and Richard Webber
© 2005 John Wiley & Sons, Ltd   ISBNs: 0-470-86413-3 (HB); 0-470-86414-1 (PB)

Introduction

At a recent launch event for the post-2001 Census update to its UK Mosaic geodemographic classification, a representative of the Experian company reported that more than 10,000 organizations in over 18 countries were using the Mosaic brand as their preferred consumer segmentation system. This is the sort of proof that geodemographics works to which Brown (1991) alludes (see Chapter 1, Section 1.1).

Yet, despite (or, perhaps, because of) commercial success, neighbourhood classification techniques have attracted criticism. It is recognized, for example, that governmental attempts to 'regenerate' poor neighbourhoods are not necessarily based on area units containing the neediest and most socially excluded members of society (Harris and Johnston, 2003). The corollary is that commercial geodemographic classifications may give misleading impressions of the sorts of lifestyles and consumer behaviours that actually take place within neighbourhoods. The charge of misrepresenting people and places is the one for which geodemographics will stand trial here.

In this chapter we begin with 'the case for the prosecution' – reviewing some of the critiques of geodemographics and neighbourhood analysis. We present what may seem like abstract theorizing and, in part, it is. Nevertheless, we urge you not to 'skip it'! By casting light on some possible weaknesses of traditional geodemographic approaches we hope not only to inform good practice (when geodemographics works well and why) but also begin revealing why some newer approaches to consumer segmentation might offer improved – or, at least, different – insights to the geographies and consumption patterns of present-day societies. In this way we set the scene for Chapter 9 where we consider some of the newer sources of geographical data and the methods by which to analyse them.

The prosecution's case primarily is built around the twin issues of the ecological fallacy and the modifiable areal unit problem (MAUP) that previously were encountered in Chapter 2, Section 2.1. It extends the argument that conventional geodemographic approaches group together members of the population not actually on the basis of their own, individual characteristics but according to some general social average for the area in which they live. The charge, 'in a nutshell', is that apparent geodemographic differences between neighbourhoods could be more a consequence of the analytical methods used to find (create) difference then a true reflection of any real-world variations between people and between the places in which they live.

In response, the defence will call witnesses to geodemographics' 'good character'. Specifically, a series of relevant case studies will be given showing examples of how geodemographics has been used and why it has been successful. Underlying the case for the defence is the contention that despite any deficiencies (and any method must have some), the enduring appeal of the neighbourhood approach is its simplicity and portability: it is readily applied and comprehended, providing a 'common currency' for organizing and comparing otherwise complex sources of georeferenced data and information. As a tool, it usefully captures and reveals some of the socio-economic, demographic, behavioural, attitudinal and consumer geographies that should inform decision making and which might otherwise remain hidden.

No final verdict is given, as that is for you, the juror, to decide! In any case, we are aware that although the court metaphor provides a useful framework for organizing the contents of this chapter, it also risks giving an unhelpful impression of setting two 'sides' against each other with the intent that one will triumph – crudely, academic theorizing versus pragmatic commercial concerns. In keeping with the spirit and rubric of this book, that is not our intent. Rather, there is much to be learned from all aspects of debate around geodemographics and so, in practice, it would be unfortunate if the 'defence' and the 'prosecution' really were as diametrically opposed as they are seemingly presented here.

8.1 The case for the prosecution

8.1.1 Trouble in the neighbourhood

Throughout this text we have tended to use the terms 'geodemographics' and 'neighbourhood analysis' interchangeably, mimicking the vernacular of the geodemographic industry. Whether we are right to synonymize in this way depends on what is meant by neighbourhood. Dictionary definitions alternate from somewhat functional explanations such as 'the immediate surroundings' or 'an area where people live', to more social perspectives such as 'a community distinct from others around it'. The trouble with the first type is its sterility – when people speak of 'their neighbourhood' they do so with a level of attachment that suggests the word cannot be divorced from the social connections and interactions that give a neighbourhood 'life'. However, the more social perspective is difficult to put into operation because the way different but closely situated people conceive their neigh-bourhood likely will vary at an individual level, with no necessary accord.

It is perhaps the knowledge that different individuals mentally map their neighbourhoods in different ways that led the two authors cited in Chapter 2, Section 2.2 to argue that 'there is no single, generalisable interpretation of their neighbourhood' (Kearns and Parkinson, 2001, p. 2103). Alternatively the statement might be verified by the number of competing geodemographic systems for sale in the marketplace! Nonetheless, it does not deny that in practice generalized, and partly standardized delimitations of neighbourhood boundaries, are defined and used – census boundary files or postal geographies are good examples.

The task of profiling neighbourhoods first involves assembling data to describe the areal objects – the neighbourhoods – to be classified (see Chapter 6). By design the boundaries of the objects may be no more than almost arbitrary lines on maps, having no (initial) existence or intrinsic meaning independent of the actual population and physical landscape that are contained within their borders. In this case it makes little sense to talk of neighbourhoods in any reified way because they are merely subjective and arbitrary population groupings. When such a 'neighbourhood' is profiled, what is actually being described and represented is the population therein. Yet, the processes of grouping populations into distinct geographical units, of profiling those units and of making strategic decisions based on those profiles will imbue the areal objects with a sense of existence in their own right – they will be assigned some degree of meaning and importance.

Repeated decision making made on the basis that two neighbourhoods are of a different kind and therefore requiring separate policy or marketing initiatives eventually could lead to neighbourhood divisions that initially were cartographic (and essentially imagined) becoming tangible and concrete. However, geodemographics should not be about constructing differences where previously none existed. Ideally, the initial design of the neighbourhood units will have a close correspondence with the existing patterning of socio-economic reality and so this knowledge of *actual* differences between neighbourhoods can well be used to alleviate, manage or market the differences in accordance with the geodemographic user's objectives.

The important question that needs to be addressed, then, is whether the boundaries of these standard (or framework) geographies in any way delimit 'natural' groupings of population? To unpack this a little, it is helpful to make the distinction between natural areas (or approximations to) that are in some sense self-defining and imposed areas that have boundaries defined independently of any (socio-economic) phenomena. O'Sullivan and Unwin (2003, p. 169) write that:

210

Imposed areas are common in GIS work that involves data about human beings. The imposed areas are a type of sampling of social reality that often is misleading in several ways. First, it may be that the imposed areas bear little relationship to underlying patterns. [. . .] Second, imposed areas are arbitrary or modifiable, and considerable ingenuity is required to demonstrate that any analysis is not an artifact of the particular boundaries that were used. This is the modifiable areal unit problem. Third, because data for area objects are often aggregations of individual level information, the danger of ecological fallacies, where we assume that a relationship apparent at a macrolevel of aggregation also exists at the microlevel, is very real.

If neighbourhood geographies are designed without consideration to any socio-economic or demographic phenomena, or to patterns of consumer behaviour that may exist, then what reason is there to suppose that classification and analysis of the neighbourhoods will reveal anything truthful, accurate or meaningful about those (hidden) patterns?

It may seem unlikely that neighbourhood geographies would be designed without giving consideration to their subsequent analytical uses. However, the prime purpose of a census is to count the population and so the main function of a census geography is to facilitate enumeration. Electoral geographies ought to facilitate a fair reflection of the population's voting behaviour within a legislative chamber (admittedly this isn't necessarily the outcome in the 'Mother of all Parliaments' but we did say ought to!). Postal geographies are intended for the efficient delivery of mail items. None of these geographies are planned for neighbourhood analysis *per se*.

Openshaw (1984, p. 3) writes:

The definition of . . . geographical objects [areas or zones] is arbitrary and (in theory) modifiable at choice; indeed, different researchers may well use different sets of units. This process of defining or creating area units would be acceptable if it were performed using a fixed set of rules, or so there is some explicitly geographical basis for them. However, there are no rules for areal aggregation, no standards, and no international convention to guide the spatial aggregation process. Quite simply, the areal units (zonal objects) used in many geographical studies are arbitrary, modifiable, and subject to the whims and fancies of whoever is doing, or did, the aggregating.

Openshaw (1984, p. 4) also argues that,

[W]hereas census data are collected for essentially non-modifiable entities (people, households) they are reported for arbitrary and modifiable areal units (enumeration districts, wards, local authorities). The principal criteria used in the definition of these units are the operational requirements of the census, local political considerations, and government administration.

Martin (1998b, pp. 109–10) contends that while formal administrative geographies may be used for planning and policy making they do not necessarily correspond to either informally or analytically identifiable social entities. Although the possibility is left open that they do (and whether they do is, as Openshaw implies, partly related to the spatial extent of the entities and the scale of thier analysis) Martin cites Morphet (1993) who demonstrates the extent to which even the lowest level 1991 UK Census enumeration district boundaries failed to fall along significant divides in urban social geography. These findings led Martin to argue that:

> [A]nalytically defined neighbourhoods can reflect many important aspects of the urban social fabric, but are hindered by the incorporation of pre-existing zone boundaries, and do not always correspond with commonly used informal neighbourhoods. For many purposes an ideal neighbourhood classification scheme would be one which could analytically reconstruct the entities recognised in informal definitions, while retaining identifiable linkages with the existing digital data which are generally available for formal geographies.

At the time the solution offered was to disaggregate census geographies using population surface-modelling techniques – a type of object-to-field data transformation mentioned in Chapter 4, Section 4.1. In scale terms, this is a top-down approach, hampered by the fact that it is extremely difficult to reverse engineer a process of aggregation with only limited, summary information of the original (non-aggregate) data still available. To prove this point try and give the definitive answer to the following question: the average of five numbers is 10 – what, then, are the five numbers? Give up? Does it help much to know that those numbers each have a value between zero and 20? Or, that the majority have a value between 5 and 15?

The additional information may indeed be helpful if it is used to model the original distribution of the five values along a number line (e.g. by estimating their variance) and that model is then used to predict the likelihood of certain values being those of the original data. However, when analysing neighbourhood data (e.g. census area statistics) usually what is available is only the sum total or the average of an atttribute value per area, with no additional information about the variance or diversity contained within (consequently, it is not obvious, for example, whether some parts of the neighbourhood contain more of an attribute such as car ownership than others). In any case, it is likely preferable (and easier) to adopt a bottom-up approach – to build zones that at least are partly optimized to the phenomena to be measured and so are in some sense self-defining. The ability to optimize zone design has been described by Openshaw as presenting a modifiable areal unit opportunity. Unfortunately, applications often (and necessarily) use pre-existing areal units such as census output areas as their building blocks (Openshaw, 1996). In such cases it is clear that the optimality of the result is as much affected by the suitability of the neighbourhood objects as the basis for analysis, as it is by the algorithm used to group them.

An important development in England and Wales has been the redesign of the 2001 Census output areas (OAs), standardized against explicitly geographical criteria. Advances in geographical information science, including methods of zone design and vast improvements in the data infrastructure have enabled the input and output geographies of the census to be separated. There are, then, two geographies: one designed for administration and enumeration – for data collection; the other, with subsequent socio-economic and geodemographic considerations clearly in mind – for data analysis.

The criteria for creating OAs include using the postal geography, specifically unit postcodes that are grouped together on a like-with-like basis, where similarity is defined by proximity and an analysis of the local housing stock. The size of OAs are constrained ideally to meet a target population of 125 households (no fewer than 40) and the aim is to delimit discrete settlements wherever possible (Martin, 2000). There are 175,434 OAs in England and Wales, of which: 37.5% contain 120–9 households; 79.6% contain 110–39 households; 5% contain 40–99 households.

The zoning process to create OAs is different from the methods (described in Chapter 6) used to cluster postcodes (or OAs) into geodemographic classes, in two ways. First, a more restricted range of census indicators was used to define the likeness of postcodes as opposed to when geodemographic types are created. Second, it would make little sense for an OA to be made up of postcodes that were miles apart and separated by postcodes belonging to different OAs, so a contiguity constraint is introduced into the zone design (postcodes that are grouped together should share a common border). Since thinking about contiguity means considering the topological relationships of the postcodes units, the consequence is an explicitly spatial technique for zone design – the location of the postcodes in relationship to each other matters. A similar claim cannot be made of the K-means method commercially used for geodemographic classification. That is a non-spatial technique since the 'distance' between (or 'closeness') of neighbourhood objects is calculated within an abstract numeric space defined by the clustering variables, not by geography.

Given the pioneering nature of the OA design in England and Wales it is undoubtedly churlish to suggest that the indicators used to characterize postcodes are a little narrow in terms of defining 'natural areas'. What we can more fairly say is that the design of 2001 Census OAs will likely decrease the population diversity within these neighbourhood units but not eliminate it. The problem remains that neighbourhood populations are frequently not as uniform as aggregate data tend to imply.

Unfortunately, aggregate neighbourhood datasets, profiles and classifications give little, if any indication of where such diversity exists and whether or not it matters. Classifications that seek to profile at a subcensus scale (e.g. by using additional sources of data available at a unit postcode level) do not entirely sidestep this problem. Although it is reasonable to suggest that unit postcodes, with about one-tenth of the population of OAs, would also have less diversity, their small size and the variable amount of data available at this scale mean that a robust classification of postal neighbourhoods requires census data to be incorporated. If it is known that 20 persons within a given OA are unemployed and also that one-quarter of all the population live in a particular postcode within the OA, then it can be estimated that one-quarter of the unemployed people (i.e. five) will be living in that postcode. However, such a method of reapportioning census data to the postcode scale assumes unemployment (or the value being estimated) is uniformly distributed across the census area – an assumption that is unlikely to be true in all cases. Again, a better estimate of the actual distribution might be obtained by population surface modelling or some other related techniques (see above, Chapter 4, Section 4.1 and also Chapter 9).

Issues of population diversity are a concern at both the neighbourhood 'building block' and the cluster levels since even if neighbourhood populations were perfectly uniform, it does not follow that all neighbourhoods assigned to a particular geodemographic cluster type are also all identical. Indeed, in a study of the problem, Voas and Williamson (2001, p. 74) concluded that 'the differences between classes [neighbourhood clusters] are generally smaller than the differences within any particular class.' If this observation generally is true then clearly it raises the risk of ecological fallacy – of assuming the apparent characteristics of a cluster type are also those of the constituent population. They might be: it all depends on how uniform the population is.

8.1.2 Is this the answer? Scale dependency and the modifiable areal unit problem

Goss (1995, p. 134) contends that geodemographics is based on the assumption that:

> these artificial statistical constructs [the neighbourhood classifications] represent individuals living in each area. In reality, of course, since the ideal of residential

sorting on the basis of uniform social identity has only been partially achieved, such analysis commits aggregation errors associated with the ecological fallacy and 'modifiable unit area problem' [sic].

A demonstration of the modifiable areal unit problem (MAUP) is provided by Table 8.1. It gives summary measures of the relationship between two 2001 Census variables for England and Wales, specifically: percentage of all persons per census unit assigned to social class AB or higher; versus the average number of cars or vans per household. Knowing that higher social class tends positively to correlate with a higher level of disposable income, it is expected that areas with a higher proportion of class AB or higher population will be associated with higher rates of car or van ownership. To put it more formally, our null hypothesis (H$_0$) is that there is *no* significant relationship between the proportion of the population assigned to the highest social classes and levels of car or van ownership, for census units in England and Wales. Our hope is statistically to disprove (falsify) the null hypothesis with a particular level of confidence so we can instead invoke the alternate hypothesis (H$_1$) – that there *is* a relationship. We expect that relationship to be positive: as levels of social class increase then so too does car ownership.

The strength of association between the two census variables is summarized in Table 8.1 by the simple but frequently used Pearson correlation coefficient (r). Calculating r to measure the level of association between social class and car/van ownership for each of the nine census

Table 8.1 Everything changes but you: the changing relationship between percentage of people in social class AB or higher, and average number of cars or vans per household for various scales of the census hierarchy in England and Wales

Census units	n (number of units for which data available)	Pearson correlation, r '% people social class AB or higher' vs 'average number of cars or vans per household'	$t = \dfrac{r\sqrt{(n-2)}}{\sqrt{(1-r^2)}}$	p (probability that correlation is due to a sampling error)
Region	9	+0.30	0.84	0.22
Unitary authorities or districts	376	+0.44	9.49	0.00
Wards	8850	+0.61	72.4	0.00

215

regions in England and Wales gives a coefficient value of $r = +0.30$. To interpret this result it is helpful to know that a perfectly positive association has $r = +1$ (meaning that as levels of higher social class rise so always do levels of car ownership by the same relative amount), whereas a perfectly negative association has $r = -1$ (as one variable increases so the other always decreases by the same relative amount) and no association has $r = 0$ (the two variables rise and fall independently of each other).

A t value also may be derived to check whether an r value represents a real association or is more likely due to a chance sampling error. Knowing that the 'degrees of freedom' is equal to $(n - 2)$, the TDIST function in Microsoft Excel, or standard statistical tables, can be used to quantify the likelihood that the apparent association is due to chance. In this case, the likelihood (probability, p) of our $r = +0.30$ being due to chance is 0.22 (or 22%). This may sound low but it is greater than the 0.01, 0.05 or – being generous – 0.10 thresholds that commonly are used as the upper limit for accepting the alternate hypothesis. Moreover (and not unrelated) the strength of association between the two variables seems itself surprisingly low, at $r = +0.30$. On the evidence available we cannot reject the null hypothesis at anything greater than a '78% confidence' ($=(1 - 0.22) \times 100$). Our results do not give sufficient evidence to support our expectation that levels of car ownership are related to social class. Why?

A clue to the answer is found by considering what happens when the Pearson correlations are recalculated at less coarse scales of the census hierarchy: at the unitary authority/district scale; and at the level of wards. As the area of the units decreases (and therefore their number increases), so the r value increases: first to $+0.44$ for census unitary authorities and districts; then to $+0.61$ for census wards. In both cases it is near to certain that the association is real. It follows that the value of r is scale dependent – the result depends on the (population) size of the area unit used to obtain the result.

Paraphrasing the last sentence gives the statement 'the value of the result obtained depends on *how* the result was obtained' and, if not a tautology, it is certainly a truism – any experimental result is dependent on the set of methods used to sample and analyse the data. Such issues are not restricted to the geographical sciences. However, there is a problem of definition that is more peculiarly geographical. Think back to Chapter 2 where boundary files were identified as a usual prerequisite for mapping neighbourhood data. There it was shown how a series of vector objects (points and lines) could be used to delimit – to form the outer edge of – a discrete, area object (in that case, the province of Labrador and Newfoundland in Canada). The perimeter defined, with apparent certainty, the location and geographical extent of the area object (the province). Any location or population group falling within the perimeter is said to belong to the province; any falling without does not.

We have already questioned how often socio-economic distributions and the geographies of consumer behaviour do exactly coincide with governmental or administrative regions – with the geography of geodemographics. If the answer is 'not often!' then apparent differences between people and places can be more an artefact of the area units used than an accurate reflection of real-world population patterns and distributions. While it is very easy to talk about geographical objects like neighbourhoods, cities, hills or forests as though their boundaries were crisp, absolute and obvious to all, the reality of most situations is rather more muddled and confused, leading to inherent subjectivity and issues of uncertainty regarding how those objects are conceptualized, modelled and represented.

Worse, any real-world patterns actually can be obscured by inappropriate design of area units. Figure 8.1 illustrates a situation where parts of

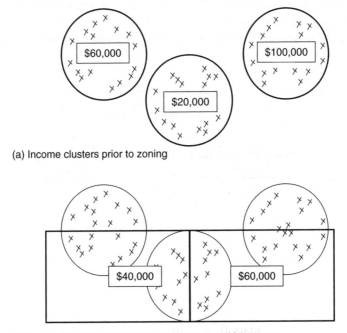

(a) Income clusters prior to zoning

(b) Income clusters bisected by and aggregated into zones

Figure 8.1 A demonstration of the zoning problem. Refer to text for explanation

three 'natural' community clusters – two relatively wealthy (with average household incomes of $60,000 and $100,000, respectively), one less so – are grouped together into two geodemographic zones. Calculating the average incomes for these two new neighbourhood objects yields values of $40,000 and $60,000, respectively – values that are not especially representative of either of the communities. It follows that there are two components to the modifiable areal unit problem: the effects of scale and the effects of zoning. Both components affect geodemographic types of analysis. Indeed, they affect all types of spatial analysis (cf. Chapter 9, Section 9.2.2).

As it happens, the r values shown in Table 8.1 are unusual. It is more common for the strength of association to increase with aggregation, not decrease at the more coarse geographical scales. The reason for the more common occurrence is that as populations are grouped into larger and larger area objects so the process tends to smooth or average out the more unusual variations or freak sampling errors found within a group. But here the aggregation effects are more closely linked to a lack of sensitivity to geographical context. Our use of the Pearson correlation coefficient assumes that any association between the two census variables can be adequately summarized by a single, linear relationship: as one goes up, so does the other; or, as one goes up, the other goes down. However, it is entirely possible that the two variables do not have such a singular relationship. Consider, for example, Figure 8.2 where two lines are shown,

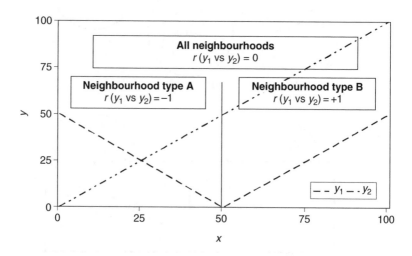

Figure 8.2 In this example the relationship between two hypothetical census variables is very different for two types of neighbourhood and, when considered *en masse*, appears not to exist at all

the first defined by the equation $y_1 = \sqrt{(x - 50)^2}$, the second by equation $y_2 = x$. The presence of x in both equations suggests that the two lines are related. They are: $y_1 = \sqrt{(y_2 - 50)^2}$.

The detail of the equations is not especially important. What is at issue is that despite their joint relationship, if the values of y_1 and y_2 are calculated in the range $0 \leq x \leq 100$ and a Pearson correlation fitted, then the result will be a coefficient value of $r = 0$, indicating an apparent lack of association between the two lines (see Figure 8.2). However, if it were known that threshold value $x = 50$ is the breakpoint – an important value differentiating two types of neighbourhood – then two coefficients might be calculated: the first for values at or below 50 (characteristic of neighbourhood type A in Figure 8.2); the second for values above 50 (characteristic of neighbourhood type B). It is now correctly found that there is a relationship between y_1 and y_2 but the nature of that relationship differs by neighbourhood. The Pearson correlation is actually $r = -1$ for $x \leq 50$ (type A neighbourhoods) and $r = +1$ for $x \geq 50$ (type B neighbourhoods).

8.1.3 Considering the geo in geodemographics

What has all the number crunching of the preceding section got to do with geodemographics and neighbourhood analysis? Are these statistical notions really relevant in practice? The answer to the second question is 'yes!' because an understanding of the importance of geographical context is essential for the effective use of geodemographic approaches to target accurately particular types of neighbourhood or particular groups of people. This is true whether area classification is used to guide neighbourhood regeneration spending or to help plan the location of convenience stores. Had we thought carefully about regional differences prior to calculating the Pearson correlations between the two census variables then we might have realized that levels of higher social class are a poor indicator of car ownership within London if it happens that those described as being of the higher social classes are also those best able to afford properties towards the centre of the city where journeying by foot, 'the tube' or by bus is often easiest and so less private transport is required.

Figure 8.3 highlights that the relationship between social class and car ownership differs for London from other regions in England and Wales, and it is that difference which is affecting the original set of Pearson correlations shown in Table 8.1. Recalculating those correlations but this time

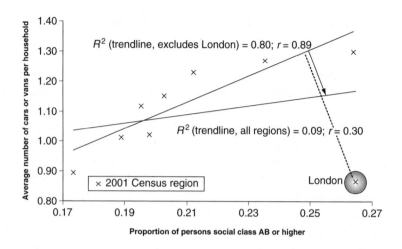

Figure 8.3 The importance of geographical context – in this example the relationship between car/van ownership and social class is very different for London than it is the rest of the country

excluding London reveals the strength of association between social class and car ownership that originally we anticipated: $r = +0.72$ at the ward level; $r = +0.75$ at the unitary authority/district level; and $r = +0.89$ at the regional level. The 'London effect' has been removed and the results conform to the more usual situation of r increasing with aggregation.

'But!' the defence may object, 'in both name and application geodemographics recognizes that geography is important. Geodemographic classifications are geographical because they attempt to differentiate different types of people with respect to the places in which those people live.'

To some extent the defence is right. Geodemographics *is* geographical in the sense that it takes neighbourhood objects – usually mail delivery or census zones – and groups them together on a like-with-like basis, based on the demographic and socio-economic profile of those objects. It is also geographical in that the geodemographic clusters, which emerge from the grouping process, usually display a distinct (apparently neither random nor uniform) pattern when displayed on a map. Finally, it is geographical in the sense that a geodemographic classification can be used to sort other geographical information – such as customer lists or the home address of students applying to universities – into geodemographic groups based on the georeference (e.g. ZIP+4 or postcode) attached to that information.

Nevertheless, there is a weak link in this otherwise geographical chain. The data-mining methods used to group the objects – procedures like

K-means or top-down, agglomerative methods (see Chapter 6, Section 6.5) – are *not* themselves geographical. In fact, many of them are inherently non-spatial. We spot a problem here. If, as a number of authors have claimed, 'spatial is special' then the sorts of methods used to handle geographical information will need to consider the geography of the problem carefully. That geographical context should not be ignored.

To appreciate the distinction between a spatial and a non-spatial grouping of geographical objects, consider Table 8.2. Grouping into two classes the nine hypothetical objects shown in the table using the closest that can be reached to a like-with-like sorting, gives rise to the following solution. Objects 1, 2, 3, 4 and 7 are grouped into one class containing a mixture of white and light grey colours. The remaining objects, 5, 6, 8 and 9 are grouped to a class containing black and dark grey colours.

The problem with this solution is it discards much of the information contained in Table 8.2. Specifically, the objects were grouped together only according to their attribute values – their colours. Figure 8.4 gives consideration as well to the locational information (the grid coordinates *x* and *y*) and, by doing so, suggests an alternate solution. In reaching the solution, each geographical object has been looked at with regards both to its own attributes but also – and critically – in regards to the attributes of the objects around it. With an eye on the bigger picture it has been decided that the difference between dark grey and white (in the case of group A) or between light grey and black (group B) is not always that great when considered in geographical context.

The point to be made here is that it could be preferable to consider the attributes of a particular neighbourhood not in isolation but with respect to the neighbourhoods around it, constructing a geodemographic

Table 8.2 Showing the results of grouping the observations – non-geographical (solution 1) and geographical (solution 2)

ID	x	y	Attribute	Solution 1	Solution 2
1	1	2	White	A	A
2	4.5	8	Light grey	A	B
3	1.5	1.8	White	A	A
4	2	2.8	White	A	A
5	1.5	2.5	Dark grey	B	A
6	4.5	8.8	Dark grey	B	B
7	2	2	Light grey	A	A
8	4.5	7.4	Black	B	B
9	4	8	Black	B	B

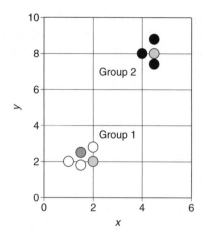

Figure 8.4 A geographical grouping of the data shown in Table 8.2

classification accordingly. Such an occasion is illustrated by Figures 8.5(a) and (b). In Figure 8.5(a) the neighbourhoods appear 'of the same colour' and it seems sensible to attach to them the same deprivation score. But, when looked at in relation to other neighbourhoods around them – in context (Figure 8.5(b)) – then it is doubtful whether the effects of deprivation or social exclusion will be the same in A, B and C. Should, therefore, their neighbourhood profile actually be the same? The answer depends on the ways neighbourhoods 'interact' with surrounding areas and at different scales, and how those varying relationships contribute to the causes and consequences of deprivation at particular localities. Consideration of contextual effects may also be relevant to the analysis of consumer behaviour, and for aspects of business and service planning.

An explicitly geographic method of building a neighbourhood profile would recognize that the ways differences between neighbourhoods are expressed and therefore how 'alikeness' is calculated is context dependent and could vary from place to place. This way of thinking acknowledges the situation shown in Figure 8.2 – that the relationships between data variables used to build geodemographic models might vary across geographical space. It is not without precedent. The concentric rings method used by Experian (and described in Chapter 6, Section 6.1) that aggregates census data at fixed distances out from a central zone, was designed to highlight regional differences in zones that appear the same at the areal unit level. And some geodemographic classifications are built by clustering London neighbourhoods separately from others on the grounds

(a)

(b)

(c)

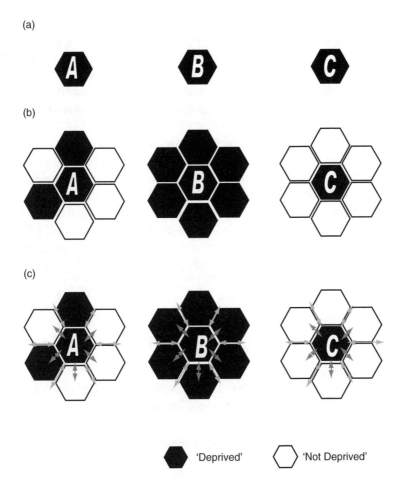

'Deprived'   'Not Deprived'

Figure 8.5 Considering context: will the effects of deprivation be the same in A, B and C? Should their neighbourhood profile be the same?

that London is very different. The issue is whether regional and other contextual differences exist outside London and also need be considered.

More generally, many conventional, GIS-based approaches for handling and aggregating geographical information give little direct consideration to the importance of geographical context. However, in the field of spatial statistics, more geographical approaches have been developed. These include various methods of analysing spatial point patterns (Diggle, 2003), multilevel analysis (Goldstein, 2003) and geographically weighted regression (Fotheringham, Brunsdon and Charlton, 2002) – see Chapter 9. These approaches are often used to provide evidence of what can be described as a collective spatial effect – the idea that people 'inhabiting' the same spaces generate some sort of collective identity or function, to the extent that the function exerts an effect upon the properties of the individual (Macintyre, 1997, pp. 1–2). This process can be understood as being due to local interaction effects and leads to second-order spatial variations, which are differences between places that are due to the shared or group effects of the population.

As the name suggests, second-order variations are different from first-order ones, which are due to changes in the underlying properties of the local 'environment' or population. Imagine a functional relationship in which the likelihood (p_i) of a person buying a prestige car is proportional to their wealth (z_i). Further imagine a location (j) with an above-average number of wealthy residents. The first-order (or compositional) effect is summarized by the relationship $p_i \propto z_i$ and from this we expect the estimated total demand ($\hat{D}_j$) for the vehicle in area j to be proportional to the general wealth of the inhabitants:

$$\hat{D}_j \propto \sum_{i=1}^{n_j} z_{ij} \tag{8.1}$$

(n_j denotes the number of residents in area j)

However, as the drivers talk proudly about their new cars and their neighbours begin to covet a similar vehicle then a second-order, collective effect emerges and – critically – may actually increase the actual demand for the vehicle in neighbourhood j over and above that originally expected:

$$D_j > \hat{D}_j \tag{8.2}$$

The collective effect serves to raise the likelihood (p_i) that any one individual will buy the type of car. Conceptually:

$$p_i = p_{1i} + p_{2ij} \tag{8.3}$$

where p_{1i} is the first-order likelihood that an individual i will purchase the car, regardless of what others around them are doing and p_{2ij} is the second-order effect on i of the collective behaviour in area j; it is a genuine neighbourhood effect. Matters can be made more complex by arguing that geographical context is also important. There may be regional differences in both the individual likelihood of buying the car and the importance of collective effects (e.g. some areas of a country can be more community based than others). It is also interesting to observe that the value of p_{2ij} is unlikely to be static – as more people buy the car it can become more likely that those who have not yet done so, will (within limits). Yet, for all the spatial complexity that could be introduced, a simple message will suffice – geography matters! The importance of remembering this is true not only for geodemographic methods but for all aspects of handling geographical information.

The prosecution rests.

8.2 The case for the defence

8.2.1 Be pragmatic!

The prosecution has put an eloquent case. It is unarguable that there are a number of questions that should be asked about geodemographics. A number of theoretical doubts have been raised.

However, the defence has approached this case from a different point of view – the commercial perspective. Put simply, it is this: geodemographics has been used in the business sector for 25 years now (confounding early pundits, who predicted a five-year lifespan at most) and it is still here, stronger than ever! Given the nature of business decisions, the cost of using geodemographics would not be borne if the technique could not prove its worth. Many major companies have used geodemographics, in its various forms, for a very long time and continue to do so. 'Something better' has still not arrived.

For examples of high-profile British and international firms that have used geodemographics to target door-drops (that is, fliers, coupons and product samples delivered door to door) see Table 8.3. This is merely a selection of examples which have appeared lately in the UK trade press.

Further evidence of the ongoing usage of geodemographics in the commercial sector has been solicited from members of the Demographics User Group (DUG), an organization that comprises some 15 large organizations in Britain and which was founded to represent to government the needs of commercial users of its demographic statistics. Table 8.4 shows how seven of the 15 members of DUG and three additional organizations use geodemographics.

The defence adopts a pragmatic stance. It is not to say that business users are totally disinterested in theory; rather that their acid test is 'does it work' (or, more precisely, 'does it work for me?'). This pragmatism probably can be seen in sharpest relief in the retail sector, where geodemographics has

Table 8.3 Companies using geodemographics to target 'door-drops' in the UK

Groups/affiliation	Brand	Product
Brighthouse	Brighthouse	Retailer
Camelot	Lotto	Lottery
Centura Foods	Bisto	Gravy mix
Centura Foods	Paxo	Stuffing mix
Colgate-Palmolive	Colgate	Toothpaste
Reckitt Benckiser	Calgon	Calcium eliminator
Jacobs Bakery Ltd	Jacobs Cream Crackers	Biscuits
Lever Fabergé	Arctic Breeze	Air freshener
Lever Fabergé	Persil	Washing detergent
Marks and Spencer	Marks and Spencer	Luggage
Nestlé	Branston Smooth	Pickle
Nestlé	Crosse and Blackwell Snackstop	Snack foods
Ocado Ltd	Ocado	Online shopping
Reckitt Benckiser	Airwick	Air freshener
Seasons Holidays PLC	Seasons Holiday	Holidays
TUI	Thomas Cook	Holidays
Tussauds Group	Thorpe Park and Chessington World of Adventures	Theme parks
Unilever Bestfoods	Flora Proactiv	Spread
Unilever Bestfoods	Bertolli	Olive oil
Virgin Group	Virgin Holidays	Holidays
Warner	Warner Holidays	Holidays

found its most secure home. It is obvious why geodemographics and retail make good bedfellows – retailers are wedded to geography. Catchment areas, or trading areas, are crucial to store performance, because catchment areas are those 'chunks' of geography from which a store's business is drawn. Geodemographics can quantify the potential for the store's goods or services within its catchment area (see Chapters 4 and 5).

The pragmatic approach taken by retail management is exemplified by an article by Penny and Broom (1988) in which they argue that the culture of retail decision making is especially receptive to answers and advice but does not appreciate esoteric analytical problems, which invariable surface in any detailed research and modelling approach. Such sentiments do not only apply to retail management. Indeed, the pressures on managers in any type of business are such these days that 'esoteric analytical problems' are probably less likely to occupy their thoughts in 2004 than they would have in 1988! They simply want to know that the solution in question, whatever it may be, demonstrably works, can be understood and is cost-effective.

Table 8.4 Applications of geodemographics by various organizations, including members of the Demographics User Group

Organization	Customer recruitment			Customer management				Retailing		
	Prospect mailings	Door-to-door distribution	Mailings to existing customers	Risk management	Pricing	Promotional offers	Channels of communication	New store location	Dealer/ branch evaluation	Dealer/store format/ merchandise
Argos		Y	Y					Y	Y	
Barclays	Y	Y						Y	Y	
Boots	Y	Y						Y	Y	Y
M&S	Y							Y	Y	Y
M&S Money	Y		Y	Y		Y		Y	Y	Y
Saga	Y		Y						Y	
Whitbread	Y	Y	Y			Y		Y	Y	Y
Honda	Y		Y				Y	Y	Y	Y
Lloyds/TSB	Y		Y	Y		Y	Y	Y		Y
csma	Y		Y		Y	Y				Y

Another feature of business people is their search for competitive advantage. The very nature of business means that if they find a technique that works for them then they tend not to want their competitors to find out about it! In the event that they cannot get exclusivity on a technique (like geodemographics) then they will want to keep secret the way that they have used it and the returns it has delivered. So, sadly, one of the best ways of proving the worth of geodemographics to doubters, an attributable case history quoting results, is most unlikely to be available for publication (disproving the adage that history is always written by the winners!). What evidence there is tends to be anecdotal or descriptive. Four case histories follow but it is interesting to note that they largely detail public-sector applications (even though the providers of the case histories are commercial providers of geodemographic solutions). Many major companies use geodemographics (as evidenced by Tables 8.4 and 8.5) but there is a dearth of published information on the detailed results of such use. Perhaps we could view these following case histories as 'witnesses' to the good character of geodemographics?

'Witness' 1: Example of the Nottingham Youth Justice Board

Edited from a report by Webber and Williamson (2004) and based on an analysis for the Youth Justice Board of young offenders' offences in the County of Nottinghamshire.

The study sought to provide evidence of the extent to which the level and pattern of youth offending varies between different types of neighbourhood. Studies from the past 30 years consistently have shown the extent to which social disadvantage is disproportionately concentrated in a limited number of areas with high levels of deprivation (Anderson *et al.*, 2001; Farrington, 2002), leading to the conclusion that area-based programmes will complement programmes targeted at individuals and households in raising the life opportunity of Britain's most disadvantaged citizens.

It is only recently that studies have started to quantify the extent to which high incidences of crime and social disorder are associated with high levels of social disadvantage. Recent evidence suggests that the variation in crime and disorder between affluent and disadvantaged neighbourhoods is considerable, indeed even greater in degree than differences in the variations of unemployment, income, poor health and educational attainment (Webber and Williamson, 2004).

Variations in the level of social disadvantage in different areas can be measured in a number of different ways. One popular approach involves the ranking of administrative wards on a composite index of social disadvantage constructed from a number of individual indicators of disadvantage. This method places each UK ward on a continuum from least to most deprived. Many social programmes are targeted at wards in the 'worst' 10% or 20% on this composite score. This study differed by using a geodemographic approach – using the 61 Mosaic types and 11 groups as the basis for analysis. It proposed that socially disadvantaged areas differ in terms of their pathology of social disadvantage, as well as in terms of their level of disadvantage; that there exist qualitatively different types of disadvantaged areas; that their different forms of social disadvantage originate from significantly different historical trajectories; and that the different types of neighbourhood are often suited to quite different types of priority area programme. The aim was to build a picture of the relationships between neighbourhood and young offenders.

The database from which the analyses were undertaken contained information on 33,905 offences, which were notified to the Nottinghamshire constabulary during the period 1 January 1999 to 7 June 2003. Information about these offences was linked to a separate file containing more detailed information about the 12,879 offenders who were apprehended in relation to these offences. 'Data cleaning' took place, removing records where the postcode of the offender was either not recorded or recognized, where the postcode of the offender was characterized by institutional rather than private residential addresses and where the offenders were apprehended within Nottinghamshire but resident outside the county. The home postcodes of the remaining 12,310 offenders were then coded with the UK Mosaic neighbourhood classification.

The distribution of offenders by Mosaic categories was compared against a baseline distribution of the estimated number of households within Nottinghamshire as of mid-2003. One Mosaic type, although present in Nottinghamshire, had no offenders: 'Global Connections'. The total number of offences per neighbourhood type was expressed as a rate per 1000 Nottinghamshire households in the same type. For the county as a whole, the ratio of offences per 1000 households stood at 74. This county average concealed a variation from 288 offences per 1000 in the Mosaic type 'Families on Benefits' to 2.2 offences per 1000 in the Mosaic type 'Sepia Memories'. The offence rates were then indexed against a value of 100 which represents county average offence rate. On this basis the offence rate for 'Families on Benefits' is expressed as 390 (3.9 times the county average of 100) and the offence rate for 'Sepia Memories' is expressed as 3 (0.03 times the county average of 100).

For each of the Mosaic types, index values were also constructed for each of 158 other ways of categorizing offences or offenders based on: level of offending (total numbers of offences, offenders); single/repeat offenders; gender; ethnicity; nature of offence; outcome; context (such as whether school day or holiday); whether the offender accepted or denied occurrence; and the offender's age band at the time of the occurrence. By this means it was possible to identify what particular types of offence particularly were associated with a particular type of neighbourhood. For example, the 'Families on Benefit' type was found to contain high incidents of: incidents by offenders of mixed race (index value 656); incidents involving arson (617); incidents where the offender was subject to a permanent exclusion order from school (499); incidents involving repeat offenders (406); and incidents involving male offenders (406).

The geodemographic approach showed that youth offending in Nottinghamshire particularly is concentrated in a small number of neighbourhood types, suggesting that an area-based approach to the problem is likely to be more effective than one which is applied at an individual level but not within specific geographic areas. The study draws attention once more to the influence of the neighbourhood in youth offending and victimization, and for the need for multi-agency working to achieve improvements in the criminogenic nature of some neighbourhoods. It points to the importance of family support for young offenders, especially after family break up and transition into areas of higher risk. Therefore, attempts to increase the social capital in areas of highest risk is vital to the success of neighbourhood renewal. The analysis helps identify the neighbourhoods and offenders at highest risk, and the types of offences that are likely to occur and could be used for resource planning, performance management and for actions to reduce the number of recidivist offenders within a particular neighbourhood.

'Witness' 2: Example of Shotton Paper Company plc

Edited from source: CACI (UK). Reproduced with permission.

Part of the UPM Kymmene group, Shotton Paper Company is a leading newsprint manufacturer. With an annual capacity of 460,000 tonnes, it produces over 20% of the UK's newsprint requirements. The mill manufactures paper from a combination of renewable Sitka Spruce grown in the UK and also relies on recovered paper from household collection systems.

Shotton Paper Company wanted to increase the amount of recovered fibre from kerbside collections in Greater Manchester, Halton and Warrington. It also needed to understand levels of newsprint consumption so it could set collection targets for business planning. Shotton acknowledged that the consumption of newsprint varies across consumers and asked CACI to help it prioritize its kerbside collection resources.

Previous waste analysis had identified a strong correlation between the volume and composition of household waste and ACORN type. For Shotton, CACI took a different approach: instead of looking at what was in the bins, it focused on the consumption patterns of different types of household. This approach is more cost-effective than traditional hand-sorting waste analysis techniques and draws on more accessible consumer research.

CACI began by looking at the ACORN profiles of those consumers reading one or more of the 211 newspaper and magazine titles that have circulation figures published by the Audit Bureau of Circulation. This enabled CACI to identify those households likely to read any of the publications and allocate the circulation figures accordingly. CACI also took account of the geographical area in which some titles, especially local free newspapers, were distributed.

The next step was to focus on those local authorities with whom Shotton were working with to improve their understanding of recyclable material hotspots. The CACI data identified individual wards within the authorities where consumption was high and as such, the amount of recyclable material would also be increased. Working with David Davies and Associates, a leading waste consultancy, CACI produced improved performance measures for each authority and equipped the authorities with tools to improve their awareness and collection strategies. In the future, Shotton will be able plan similar campaigns in new locations based on the improved consumption data derived from the ACORN-led consumption analysis.

'Witness' 3: Example of The University of Central Lancashire

Edited from source: CACI (UK). Reproduced with permission.

The University of Central Lancashire traces its roots back to the Institution for the Diffusion of Knowledge, founded in Preston in 1828. In terms of student numbers, it is currently one of the top 10 universities in the UK. As part of its plans for growth the university needs to ensure that it has a full understanding of the location and characteristics of potential students.

As with any university, the number of students enquiring about degree courses is far in excess of those who apply, accept a place or actually enrol on a course. This is sometimes known as the 'recruitment funnel' as student numbers reduce steadily throughout the process. For example, out of the 45,000 enquiries received by the university, typically 20,000 turn into actual course applications.

The University of Central Lancashire felt it important to discover why over half the enquiries are not taken any further. Could it be that the wrong people are being targeted? Where do the past and present student population come from? Why do those attending the university choose it? It was hoped that an understanding of the socio-economic groupings of students attracted to the university could be achieved by a geodemographic approach and this information used for proactive targeting.

The method of analysis combined the ACORN product with InSite – CACI's geodemographic market analysis tool that has helped over 200 businesses improve their understanding of their locations and customers. Together they offered tools for spatial planning, student application analysis, catchment planning and mapping.

The university was able to identify areas where targeted student recruitment activities would have the greatest impact. The analysis was also able to identify areas where no marketing activity had taken place but where applications to the university were still popular. This allowed marketing resources to be employed in the most effective manner. For example, information on the cost of living and personal safety in Preston could be targeted to areas where applications from women are most prevalent. The combination of InSite and ACORN has meant that careers visits and local advertising can be targeted precisely, participation can be widened, knowledge of student needs can be enhanced and these needs can be matched with marketing material. The return on investment (ROI) will be measured by the increased applications to the university.

By understanding the market, the university will be able to have advanced knowledge of what information potential students in each area of the UK require in order to make their application. By providing these students with relevant and targeted information more enquires can be converted into applications and enrolment increased. The university already has further plans for detailed analysis of the alumni population of the university so that information can be gained about where graduates move to, what job they decide to do and the resulting life trajectories they experience.

'Witness' 4: Example of Camelot Group plc

Based on Sleight, Smith and Walker (2002).

Camelot Group plc won the first contract to operate the UK's National Lottery, back in 1994. The contract ran for a seven-year term, with a competitive bidding process taking place for a new contract in the first half of 2000. Camelot's Retail Services Division decided that GIS technology had moved on since the initial selection of lottery outlets had taken place in 1993. In 1998 they asked Target Market Consultancy (TMC) to review and revalidate the retail network selection methodology. At the time, Camelot operated some 35,000 terminals nationally.

The first stage of the project involved reviewing available data sources. In-house data included terminal transaction history, all relevant market research, and a database of actual and potential retail outlets where terminals were currently sited, or could be sited. External data that might be relevant to the project were also evaluated. These included shopping flow data, workplace data from the census and information-locating retail outlets relevant to Camelot – primarily supermarkets, post offices, multiple and independent CTNs (confectionery/tobacconist/newsagents), and convenience stores.

The problem of how to best configure the retail network was conceptualized as a demand and supply issue. Consumer demand for lottery purchases could be modelled using a geodemographic approach, relating demand to types of geographical area. Useful to this purpose, Camelot ran an ongoing advertising tracking study which interviewed 1000 consumers each month, identified spend by demographics and asked 'where bought' questions. Respondents' home postcodes were also captured. The physical placement of lottery terminals at specific locations represented supply points. The objective was to match supply to demand in such a fashion that supply of terminals was optimized. The simple expedient of putting lottery terminals in all available outlets (maximizing supply) would not make economic sense because of the high cost of online terminal supply and maintenance.

The recommendations were that a spatial interaction (gravity) model should be built and calibrated, taking account of flows from residential areas, workplace and shopping trips (cf. Chapter 4, Section 4.5). A GIS should be developed, containing the 'universe' of relevant retail outlets. Census of population data should form the basis of a consumer potential model for lottery playing. The gravity model initially should be developed in a relatively small geographical area, for instance the trading area of a large shopping centre and then, if successful, be rolled out to cover the country. As a part of the initial development work, candidate datasets should be evaluated.

These recommendations were accepted and Business Geographics Ltd (BGL) with GeoBusiness Solutions were chosen to develop and implement the model. The Travel-to-Work area of the town of Reading, England was chosen as the study area for model development. As part of the development process, an exit survey was designed and implemented to provide information on trip behaviour of lottery ticket purchasers (different types of trip were identified in town centre/suburban/rural locations). The process also involved the creation of a 'customized' geography (Camelot Selection Districts, CSDs, constructed from aggregations of census output zones), the development of the consumer potential model and its subsequent implementation within a GIS. The consumer potential model was based on the Camelot tracking study, which at the time of the study had nearly 24,000 respondent records available, each with a record of social grade. By modelling frequency and volume of spend by social grade and matching this information to the socio-economic groups present in the 1991 Census, a model of consumer expenditure on lottery purchase could be calculated for Camelot Selection Districts and regional trends identified.

The resulting 'Optimum' model was implemented in Camelot's offices and it quickly became a vital part of Camelot's retailer selection process. The 150 field sales executives were provided with reports and maps that indicated how many terminals could be supported in each CSD, thus providing guidance for locating new or replacement terminals to supplement the field sales executives' local knowledge. Optimum continues to be used today, and its capabilities and applications continue to be enhanced.

8.2.2 Back in the neighbourhood

Having considered some successful applications of geodemographics, let us turn to the question of defining neighbourhoods raised by the prosecution. What are 'natural neighbourhoods?' At what scale do they exist – the level of a unit postcode, perhaps, or maybe the level of a 2001 Census output area in England and Wales? Or, perhaps, it is at some other level of area unit? Is, in fact, everyone's definition of their own neighbourhood perceptually different? Certainly this is interesting to ponder. Yet, again in a commercial context, there's a need for pragmatics: we use what we've got! In the early days of geodemographic development in the UK, electoral wards were used. Then, when the requirement for finer analysis came along and the data infrastructure was able to support it, census units were used. Subsequently, data at unit postcode level became available and so Mosaic was produced at that level too. Nowadays, with the new post-2001 Census classifications, the majority operate at postcode level, although some have stuck to census output areas. Geodemographic classification techniques can be applied at any of these levels, though often smaller units of area are preferred because demographic homogeneity is likely to be greater.

Notwithstanding our differences, we agree with the thrust of the prosecution's critique – that if the user misunderstands the technique, there is a danger of her or him misapplying it. Certainly, this is an area where it is incumbent on vendors of geodemographics to do their best to educate their clients into the strengths and weaknesses of the method. They should make users aware of issues such as the ecological fallacy (whether they choose to call it by that name or not!). The presumption appears to be they don't already, yet in reality the awareness exists. For example, the ACORN profiles that can be viewed at the website www.upmystreet.com come with the following warning:

> This description is intended to illustrate likely consumer preferences and behaviour and does not describe a specific locality or its residents. Please note, it may not always be entirely accurate for your postcode. You should not base important decision making on the ACORN classification alone.

The prosecution is from an academic background and, as such, perhaps does not fully understand the nature of business relationships. In the private/commercial sector, vendors usually work directly with end-users and there needs to be a high level of consultancy back-up and service support from those suppliers. It is a very competitive market so vendors need to fully satisfy the needs of their customers in order to retain their business over time. Long-term vendor-to-customer relationships are important to the profitability of the vendors. Ultimately, vendors are only as good as the products they supply and the way they supply them. Associated with this point, there have been many academic criticisms of 'commercial' geodemographics in the past. There seems to be a feeling that commercial authors of geodemographics have perhaps 'not done it properly'. Our experience over many years suggests that this is not at all the case – the developers give much thought to methodology, some have very extensive experience of developing classifications and, indeed, have developed sophisticated methods. The main difference between these commercial developers and their academic opposite numbers is that the former tend not to publish details of their methodologies, whereas the latter do!

To sum up, whatever the virtues of the prosecution's case, ultimately it is rather akin to accusing an orange for tasting fruity! Their argument essentially criticizes the apparent simplicity of the geodemographic approach, losing sight of the fact that it is that same simplicity which lies at the heart of its widespread appeal and ease of use. It is also downplays the merit of a classification that tries to 'let the data do the talking', to see what important socio-economic and demographic patterns emerge and to consider how they may be targeted.

It was a realization some years ago that there is little 'hard evidence' to prove geodemographics that led the Census and Geodemographics Group (formerly the Census Interest Group) of the Market Research Society to implement what became known as the Luton Case History. The study was thorough and gave good evidence that geodemographics 'worked'. To quote the chairman of the group: 'this study found that a geodemographic system worked well in explaining actual differences in product consumption patterns between neighbourhoods.' The results of that study are summarized in the following case study.

The defence rests.

Validating geodemographics – the Luton case study

Barry Leventhal, Teradata, a division of NCR; and chairman of the Market Research Society Census and Geodemographics Group.

Geodemographic classifications have been widely used in the UK for over 20 years, without any formal proof or evidence that they do actually predict how market consumption rates vary by geography.

In 1995, the Market Research Society's Census Interest Group (CIG) formed a working party to examine this issue. The team included representatives from the Group Market Research department at Whitbread plc who had been using geodemographics since the early days of the industry. The team quickly grew to include representatives from BMRB International, who operate the Target Group Index (TGI) (see Chapter 3) and two leading census agencies: CACI and Experian.

The working party started by looking at geodemographic profiles of product usage on the TGI, in order to see the how the discriminatory power and patterns vary between products. For example, many products – such as dishwasher ownership – show an 'upmarket' bias, while others – such as cigarette smoking – show 'downmarket' skew. These profiles were all based on the Midlands BARB TV region, in order to reduce the likelihood that the differences could be caused simply by regional factors. However, the fact that the profiles showed apparent differences between different products and places was no guarantee that these were also actual differences.

The group therefore wanted to test whether geodemographic profiles could really predict differences in consumption at a neighbourhood level. The only way to achieve this was to survey actual consumption rates for neighbourhoods within a town and compare the observed behaviour with the geodemographic predictions.

The town selected was Luton – a local authority district containing 340 census enumeration districts (EDs) and a rich mix of geodemographic types within a contained area. Luton was also the location of Whitbread's main offices and so was well known to their representatives on the working party. This local knowledge was useful for 'ground truthing' the results.

The test was structured using ACORN, based on the 1991 Census. Similar results could doubtless have been produced using other geodemographic classifications, and the survey results were post-analysed by both ACORN and Mosaic. Three ACORN types were selected, corresponding to 'upmarket', 'middle-market' and 'downmarket' demographics. These ACORN types are summarized as: Type 11 – Affluent working couples with mortgages, new homes; Type 31 – Home owners in older properties, younger workers; and Type 44 – Multi-occupied terraces, multi-ethnic areas.

A sample of nine census wards was drawn containing the target types and, within each ward, two target EDs were selected. This gave a sample of 18 EDs in total, six for each ACORN type. A random sample of addresses was then drawn for each ED, using Ordnance Survey's AddressPoint product (see Chapter 3) to ensure that all of the addresses actually were located in the designated neighbourhoods.

The survey was undertaken by BMRB during September 1996 and was, in some ways, equivalent to a mini-TGI survey. BMRB interviewers visited the selected addresses and conducted in-home interviews to capture demographics and media consumption. The respondents also completed a short questionnaire on product usage. By repeatedly revisiting the addresses, interviews were achieved with 87% of all eligible homes (excluding non-contacts) and a total of 870 interviews were obtained – an average of 48 per ED. This was a sufficiently large sample to analyse consumption rates at ED level and the high response rate ensured that the results were representative of the population in each ED.

Initial analysis was undertaken to look at the patterns of consumption rates by neighbourhood types, comparing the survey results with TGI profiles for the Midlands. The extent of agreement was marked and could not have been due to chance. With a collection of measurements on each survey respondent in each ED, an approach was needed that took account of the many relationships in the data. Therefore, multivariate analysis seemed an appropriate next step. At this stage Robin Flowerdew, then of the geography department at Lancaster University, joined the working party.

An initial cluster analysis was conducted, using the product consumption rates for the 18 EDs, in order to classify the neighbourhoods into three groups based on their consumption patterns. If ACORN was a good predictor of consumption, then the three consumption clusters should correspond to the three ACORN types. The results of this test are shown in Table 8.5. Each consumption cluster did correspond to a distinct ACORN type and 15 out of the 18 EDs were in perfect agreement.

However, three EDs did not agree and the working party decided that further analysis was needed in order to try to understand these outliers. An alternative 'fuzzy' approach to clustering was selected (and help was enlisted from Zhiqiang Feng of the same geography department). The fuzzy clustering approach is an extension of traditional 'hard' clustering, in which each ED is assigned degrees of membership for belonging to different clusters. Therefore, rather than forcing each ED into one cluster, the approach identifies EDs with similarities to several clusters. The ED is allowed to belong to more than one neighbourhood type and its degree of belonging to each can be calculated. A fuzzy classification of the ED consumption data showed that the three clusters were fairly distinct and did not explain the outliers which all had strong links to a single cluster, albeit in ACORN terms, the wrong one.

The next step was to create a fuzzy geodemographic classification using census data for the 18 Luton EDs, in order to see whether that would explain the outliers. A set of 44 census variables were chosen, covering mainly economic, demographic and social characteristics. Although this was not the same list of variables as used in ACORN, the fuzzy classification was

Table 8.5 Distribution of Luton study census EDs across ACORN and case-specific consumption clusters (* denotes residual case)

		ACORN type		
		44	31	11
Consumption	1	5	1*	
cluster	2	1*	4	
	3		1*	6

not very different from ACORN. The main difference was that one of the three outlier EDs turned out to have stronger links with the census cluster equivalent to ACORN type 11 – matching its consumption cluster – rather than type 31 as assigned by ACORN.

The remaining two outliers were explained in other ways. One was found to have undergone substantial change in socio-economic characteristics since the 1991 Census and the Luton Survey in 1996, due to changes in its residential structure over the five-year period. The second was examined with a final test considering whether ACORN and its fuzzy version were too general purpose for the products measured in the Luton survey. A special-purpose fuzzy geodemographic classification was created, giving higher weights to those census variables related to product consumption. Using this classifier, the remaining outlier came out with high degrees of membership in two clusters – one equivalent to its ACORN type and the other matching its consumption pattern.

In conclusion, this study found that a geodemographic system worked well in explaining actual differences in product consumption patterns between neighbourhoods. The small number of exceptions could be accounted for by either change since the census data underpinning the commercial classification and by the application of a special-purpose discriminator or by a fuzzy classification approach.

8.3 Conclusion

In this chapter we have sought to balance some theoretical weaknesses of geodemographics with some practical advantages. This reflects our desire to be candid about the merits and demerits of the method, and thereby encourage good practice. The strengths and weaknesses of any analytical approach need to be considered within the context of what the users hope to achieve with it, why and with what additional resources available to them. Some of the limits to geodemographics that we have outlined will be more relevant to some analyses, the conclusions drawn from them and the practical consequences of responding to those conclusions, than to others. The onus is on the user, the vendor and the nature of their business–client relationship to ensure that the analyses that take place and, most particularly, the inferences drawn and acted upon are responsible, reasonable and well founded. While it is tempting to suggest that geodemographics is more the tool of commerce, to do so would be to miss both its origins and recent deployments in areas of substantive public policy research.

One area of critique – criticism, really – that we have not looked at in this chapter are studies of the 'discourse' of geodemographics. These can be understood as looking at the language and nomenclature of geodemographics, and what they reveal about the underlying assumptions and practices that are involved in its marketing and use. Such studies tend to focus on the labelling of neighbourhood clusters and the ways they caricature places but, more particularly, people – understandably so (see Chapter 6, Section 6.8). They also question the extent to which the world really can be 'internalized' within a computer model and then 'manipulated' by an independent user (cf. Chapter 5, Section 5.2). Such discussions are interesting and important and we have flagged them as further reading to this chapter (they are also sometimes hard going and perhaps overly focus on the theorized and not actual business environments in which geodemographic vendors and users operate). A further area of concern – issues of data privacy – is considered in the following chapter.

Summary

- Neighbourhood classification techniques have attracted criticism from academic quarters. Some of these are more technically driven, focusing on the clustering methods used. Others question the way people and places are represented by geodemographic systems.
- The formal, administrative geographies that are used for planning and policy making may be 'imposed' on populations – they do not necessarily correspond to either informally or analytically identifiable social entities.
- However, the design of the 2001 UK Census output areas has used principles and methods of geographical information science to standardize the zones against geographical criteria and in a way that benefits subsequent geodemographic analysis.
- The modifiable area unit problem (MAUP) affects all analysis of geographic information and has two components: the effects of scale and the effects of zoning.
- Techniques such as K-means do not constrain the cluster outputs to fit geographical assumptions such as Tobler's first law of geography but, as such, may 'smooth out' contextual differences in the way clustering variables interrelate at regional or other sub-national scales.

- Nevertheless, geodemographics has been successfully used in the business sector for many years and in a wide range of applications that includes: prospect mailings; door-to-door distribution; mailings to existing customers; risk management; pricing promotional offers; channelling communications; planning new store locations; branch network evaluation; and store formatting and merchandizing.
- Despite an unfortunate dearth of published, commercial evaluations, a number of case histories and a validation exercise in Luton, England suggest that geodemographics does indeed 'work' for its users.

Further Reading

- Curry, M. (1998) *Digital Places: Living with Geographic Information Technologies*, Routledge, London.
- Goss, J. (1995) Marketing the new marketing: the strategic discourse of geodemographic information systems. In *Ground Truth: the Social Implications of Geographic Information Systems* (ed., Pickles, J.), The Guilford Press, New York, pp. 130–70.
- Openshaw, S. and Wymer, C. (1995) Classifying and regionalizing census data. In *Census Users' Handbook* (ed, Openshaw, S.), GeoInformation International, Cambridge, pp. 239–69.
- Sleight, P. (2004) *Targeting Customers: How to Use Geodemographic and Lifestyle Data in Your Business*, World Advertising Research Center, Henley-on-Thames.

9
New Data, New Approaches: from Geodemographics to Geolifestyles

Learning Objectives

In this chapter we will:

- Change our focus from neighbourhood classifications to sources of individual and household data.

- Consider how such sources of lifestyle data are of use in one-to-one and direct marketing and in building 'geolifestyle' classifications.

- Revisit the opposite of the ecological fallacy – the atomistic fallacy.

- Look at how some GIS-based and other techniques may be used to avoid the atomistic fallacy, offering 'bottom-up' methods that explore and help explain spatial patterns in lifestyle datasets.

- Look at issues of data uncertainty and personal data protection when using lifestyle datasets.

- Present the last of our expert case studies, this one entitled 'Lifestyles analysis and new approaches', written by Gordon Farquharson of Streetwise Analytics Limited.

Geodemographics, GIS and Neighbourhood Targeting   Richard Harris, Peter Sleight and Richard Webber
© 2005 John Wiley & Sons, Ltd   ISBNs: 0-470-86413-3 (HB); 0-470-86414-1 (PB)

Introduction

So far in this book the focus of our attention has been on neighbourhood classifications, by which we mean the like-with-like sorting of variously defined administrative zones into distinct types or groups. Although many of these zones, such as postal or ZIP+4 units, contain a small number of the population they always have an area or geographical component. In any case, governmental sources of geodemographic data usually are not available for these more detailed geographical units – the dissemination of most national census statistics is a case in point. In order to offer the level of precision required by users of geodemographic systems, classification builders often reapportion census data to smaller zonal divisions and supplement them with sources of personal or household data. The nature and availability of these sources were discussed in Chapter 7 and shown to vary from country to country.

The existence of these non-aggregated data raises a question: if datasets are available that are not area based, then why not use them to classify persons or households directly using techniques analogous to those described in Chapter 6? Such an approach would avoid the problems associated with the ecological fallacy, since if the 'type' of any particular person is already known then there is no need to try and infer that information from the neighbourhood instead.

In fact, at least one company does offer such an approach. CACI's People*UK product is a mix of geodemographics, lifestyle and lifestage data classifying every person on the UK electoral roll into one of 46 different types. These include: 'Silver Spoons', 'Theme Park Families', 'Loan-Loaded Lifestyles', 'Staid At Home,' 'Beer and Bookies', 'Put the Kettle On' and 'Church and Bingo' (www.caci.co.uk). CACI also has produced eTypes, a classification for understanding online consumer behaviour, with seven main categories and 23 subcategories. Apparently, 'Dot Com Dabblers' tend to be found in Yorkshire, Lancashire and the North East of the UK; 'Virtual Virgins' in Scotland (www.etypes.info).

Individual or household classifications, built from individual or household data, are of obvious use for one-to-one and direct marketing. By completing shopping surveys or extended product warrantee cards, or by holding loyalty/affinity cards, the source data are often provided and updated by the individuals/households themselves. Unless the respondents are given to lie or change their life circumstances frequently, it is likely that the information they provide will be representative of their actual living arrangements and consumer lifestyles. For the subset of the populace that provides such information it could be of more relevance to marketing and other applications than that obtained through more conventional, statistical sources.

A second question follows: how large is the subset? Acxiom (Claritas UK) promotes its Lifestyle Universe product as having, at its core, data received from three in four UK households, as including 380 pieces of detailed information on 44 million UK individuals and as being updated quarterly (www.claritas.co.uk). It is easy to understand why such products increasingly offer more appeal to marketers than decennial census updates covering, by comparison, only a narrow range of geo-demographic data. (Also why governments are considering using new datasets to avoid what they often perceive to be a costly process of census enumeration – see census case study, Chapter 7.)

Individual, household or small-area classifications based on lifestyle data are sometimes known as geolifestyle classifications. The geo gives a connection with more 'traditional' geodemographic approaches but, aside from this, the use of the prefix might be questioned. It was observed in Chapter 8 that while neighbourhood classifications are built using geographical data, aggregated into geographical units, the methods of clustering rarely are geographical in themselves (the K-means method, for example, is not). There are advantages in not imposing *a priori* geographical assumptions upon the clustering routines but, instead, attempting to let 'the data do the talking'. However, it is also somewhat tenuous to imply that individual/household classifications are especially geographical if the geography really only emerges once the results are plotted on a map. Indeed, there is a risk of committing what is essentially the opposite of the ecological fallacy: the atomistic fallacy, understood as a too myopic focus on individuals or households at the expense of ignoring the larger social, economic and cultural contexts in which those individuals and households are situated (cf. Chapter 8, Section 8.1).

In this chapter, when we talk of geolifestyle classification we mean to specifically consider the results of using explicitly spatial techniques to draw out geographical attribute patterns from lifestyle types of database. This conception is more aligned to developments in spatial data mining, point pattern analysis and the repeat-testing methods of data exploration from the field of geocomputation (Longley *et al.*, 1998; Miller, Han and Fraser, 2001; Openshaw and Abrahart, 2000); less aligned to transferring conventional segmentation techniques to new sources of data. We offer some 'hows and wherefores' of applying simple and, initially, GIS-based techniques of spatial analysis to lifestyles data, before going on to look at some potential pitfalls, including uncertainties with the data and issues of personal privacy. We begin, though, with an overview of using lifestyle data in marketing, in a case study that raises a number of issues to be discussed further in the remainder of the chapter.

Case study: Lifestyles analysis and new approaches

Gordon Farquharson, Streetwise Analytics Limited

Before we start exploring its applications, it is worth attempting to define what the term 'lifestyle data' really means. Unfortunately there are few, if any, authoritative references devoted to this type of information. In general, lifestyle data is an umbrella term that covers information collected about consumer choices and behaviours. By choices we mean product ownership, interests, pastimes, shopping preferences and so on. Lifestyle surveys might then question how a consumer has reacted to marketing 'noise'. Behavioural questions, on the other hand, illustrate how a consumer interacts with the market. For example, whether a consumer carries a credit card could be regarded as a reaction or choice; whether they pay their monthly credit card balance in full is an interaction or behaviour.

So, we are describing information obtained from a type of survey that asks consumers how they live their life. Why is this so important when market research companies have been doing surveys like this for years? The real value behind lifestyle data and why they recently have come to prominence in the marketing arena is the sheer volume of consumers providing information. Through dedicated surveys, lifestyle data companies collect millions of responses every year, making this type of information a prime resource. Throw in millions of additional responses per year from the collection of extended product registration cards for brown and white goods and you can understand why the direct marketing industry is a thriving marketplace.

Lifestyle questionnaires specifically ask consumers to consent to receiving other marketing communications and so lifestyle data primarily lend themselves to direct mailing applications. Being able to target consumers who are known to be interested in golf rather than targeting neighbourhoods with an 'unusually' high prevalence of golfers should result in superior response to a direct marketing campaign among those targeted. The traditional, neighbourhood approach may, however, be better for reaching 'new' golfers – that is, people who play golf but for whom lifestyle data are not available. Alternatively, the lifestyle database can be extended by the modelling of lifestyles onto the portion of the population where it has not been possible to collect information. This brings its application into a wider universe, albeit with reduced expectations. In the direct marketing sector this proposition is becoming common with Claritas (The Lifestyle Universe), CACI (Lifestyles*UK) and most recently, Consodata (OmniLifestyle, www.consodata.co.uk) promoting lifestyle databases with full national coverage.

From the actual, non-modelled responses it is also possible to identify geographic differences and preferences. For example, just by plotting on a map the locations of households who shop at a particular store, it is easy to identify how far the attraction to the store extends. While this kind of catchment modelling still needs some development it does present local insight that is difficult to measure through any other data source. Extension of this application is relatively straightforward through controlled weighting and modelling to generate more robust estimates at a local level.

A further extension of this local data is in the development of so-called 'geolifestyle' classifications. Similar to geodemographics, geolifestyle segmentation focuses on using the survey responses as the inputs into a cluster analysis model. The responses to each question can be aggregated to provide percentage estimates for small-areas units. The result is a segmentation solution that discriminates each unit based on lifestyle choices and behaviours. Understandably, the effectiveness of these solutions will vary considerably because they are very sensitive to the amount and the age of data available.

This variability highlights a limitation when using lifestyle data. Although response rates do vary across different socio-economic groups, the underlying UK rate is approximately 6% for a typical postal survey. And, even when we consider a composite file of responses accumulated over a number of years, there is still likely to be a number of significant 'holes' in the database. We have already seen that models can be built from the actual data to fill these holes. This can tempt us to believe that just because it is possible to build a model at individual or household level it will necessarily provide the best solution. It may be better and is generally easy to aggregate to postcode or other geographies. However, such approaches can yield poor and usually flat 'solutions'. Both individual/household and postcode classifications can dilute the potential benefits of the underlying data. Even for true individual or household applications, many of the solutions are built as national models that do not readily incorporate the local effects and contexts that influence certain types of choices and behaviours.

The other main criticism of using model-based methods to extend lifestyle databases is that the base surveys provide a biased view of consumer worlds. It is certainly true that a typical response profile is not representative of the populace as a whole. The data companies do not dispute this but point out that the vast number of questionnaires involved should lead to the risk caused by this bias to be minimal for many applications. What is important is the level of care taken to model the rest of the data effectively, where it has an important bearing on any inferences.

This leads us to query what can be done to utilize these rich sources of (lifestyle) data in such a way that inferences made from them are robust and actionable. For both geographical and individual based applications the answer involves the optimal integration of geography into our models. This does not mean simply adding a geographic variable into our usual model and treating it in the same way as the others but instead using techniques which assist in generating models to capture 'neighbour-hood effects' detected in the attributes of the lifestyle data.

Two of the more prevalent of these techniques are multilevel modelling and geographic fusion. The former includes geography as a stratum and forms an integrated, hierarchical model with error terms and measures of variance included for each level. By nesting, for example, households within postal units, within census zones, within regions, the model can quantify the significance and divergence of different types of lifestyle at different geographical scales. Multilevel modelling is a powerful technique and realizes robust local models. Geographic fusion takes a different approach and uses donor imputation techniques, the most common known as hotdecking (used in the development of OmniLifestyle), to generate observations within a predetermined geography. Because hotdecking techniques use some combination of neighbouring values to fill missing ones, the models are sensitive to biased data but generally lead to robust estimates with no loss of consistency across variables (i.e. the covariance matrix is preserved).

Most current techniques are still straitjacketed by the geographies within which they are applied. To advance the quality of estimates derived from lifestyle data it is essential to maximize the utility and flexibility of the data by developing an optimal geography for modelling. Key to this step forward is to recognize boundaries where an acceptable statistical difference exists. These boundaries can then form the reporting geographies, ensuring that the lifestyle data are represented with maximum effect and more optimally incorporated within multilevel modelling or geographic methods.

In summary, lifestyle data are viewed as an important ingredient into many marketing plans. While the response rates are always challenging, lifestyle companies are exploring new channels for collecting information.

Internet surveys are increasingly common and appeal to segments that were traditionally resistant to volunteering lifestyle information. Other channels are emerging as suitable media for collecting information, particularly the telephone, where DataLocator (www.datalocator.com) has built a successful programme. The continued success and evolution of lifestyle datasets will depend on the quality of the data collected and, more importantly, the quality of the inferences made from them. Appreciation of the benefits and statistical limits remain central to effective consumer insight.

9.1 Using GIS to map lifestyle data

The preceding section identified that lifestyle data can be used to identify and analyse local patterns and trends. A straightforward means of doing this is to plot the data on a map. Each lifestyle record usually has an address field and this residential location can be given a point grid coordinate using products such as Ordnance Survey UK's Address- and Code-Point products (see Chapter 3, Section 3.1). Having done so, it is easily visualized using a GIS. In this way, Figure 9.1 shows a subset of a lifestyle dataset, indicating the locations of survey respondents who identified themselves to be 'single' in a particular town within southern England. Each point is at the population-weighted centre of a unit postcode (the centroid) and the size of the symbol is proportional to the number of 'single' respondents per postcode.

Given Figure 9.1, a question reasonably asked is whether 'singles households' geographically are concentrated into certain parts of the town. In previous chapters the way to answer this would be to link the records to the neighbourhood types in which the households are located. Although this would be a predominantly automated procedure in the sort of geodemographic information systems described in Chapter 5, it was also shown, in Chapter 4 (Figure 4.8), to be equivalent to overlaying a neighbourhood objects layer upon the points layer in a GIS, undertaking a point-in-polygon analysis and using this to group together the points found to be located in each of the neighbourhood classes. As such, the conventional geodemographic approach can be regarded as a 'top-down' method whereby the preconstructed neighbourhood classes would be placed atop of and down on the lifestyle data for analysis. What we are seeking here is a 'bottom-up' approach. By this we mean to draw patterns out from the lifestyles data, defining the spatial extent of any patterning on the basis of what we find, as opposed to fitting it to some *a priori* classification of neighbourhood types.

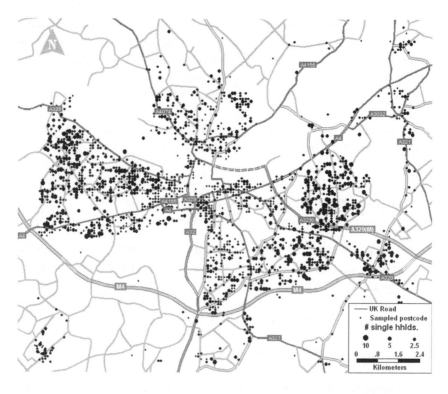

Figure 9.1 Distribution of lifestyle surveys respondents classifying themselves as 'single' for a settlement in southern England

Simply 'eyeballing' Figure 9.1 suggests locations where clusters of singles households appear to be found. However, there is a need for caution and not to jump to the conclusion that any apparent patterns are necessarily significant. Their significance is unproven. You can probably see patterns in Figure 9.2, too, but all we have done for this map is take the same set of locations as Figure 9.1 and *randomly* reassign the original attribute data to an alternative one of the locations. Any apparent patterns in Figure 9.2 are due to the geography of the study region and the geography of the survey response, not the actual residential choices made by singles households.

9.1.1 Grid-based analysis

The issue of assigning significance to apparent clusters of a lifestyle attribute is returned to in Section 9.2. A more immediate problem is that when we look at each point on Figure 9.1, like is not actually compared

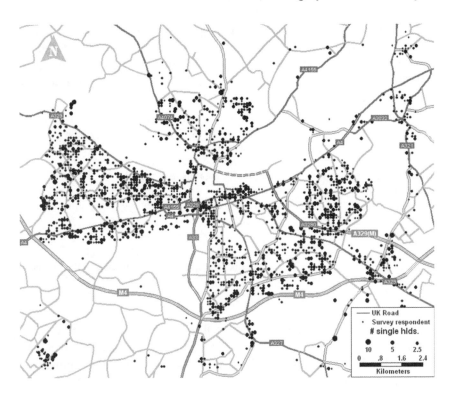

Figure 9.2 Map showing the random redistribution of the 'singles' attribute values to alternative locations in the same study region as Figure 9.1

with like. This is because the number of survey respondents in each postcode varies (from 1 to 16; the mean average is 1.9, the median, 1). This variation may reflect the fact that certain subsets of the population are more likely to provide lifestyle information than others but is also due to unit postcodes being of varied population size.

It is necessary, therefore, to standardize the local counts of singles households with respect to the local count of survey respondents. The proportion of the respondents that class themselves as single could then be calculated for each unit postcode. However, with averages of only 1–1.9 respondents per postcode, small-number effects would emerge. The averages suggest the proportions would be constrained to a non-continuous range of values and approximate to one of three discrete values: 0, 0.5 or 1. In other words, the data do not support robust estimates of the actual (real-world) proportions of singles households at the unit postcode scale.

The problem is that the postcodes are not of a large enough population size to permit proportions robustly to be calculated from the lifestyle sample available. An alternative approach is illustrated by Figure 9.3. This adopts a quadrat-like approach and aggregates the point data into a regular grid to calculate the proportion of 'single' respondents per grid cell. Grids are easy to define (cf. Chapter 4, Section 4.1). The following text specifies the grid shown in Figure 9.3:

NCOLS 20	<number of columns>
NROWS 20	<number of rows>
xllcorner 465212	<lower left x coordinate of the grid>
yllcorner 166713	<lower left y coordinate of the grid>
cellsize 700	<cell length>
nodata_value −9999	<value for indicating an absence of data within a cell>

1	2	3	4	5	6	...	19	20
21	22	23	24	25	26	...	29	30
...				<attribute data: one item (in this case an ID) per cell>				
...								
391	392	393	394	395	396	...	399	400

Having derived proportions per grid cell the information can be used to fill missing values in the lifestyle database. For example, imagine no information has been collected for a postcode located in the highlighted grid cell found near the centre of Figure 9.3. An estimate of the proportion of singles households to be found in that postcode is provided by the cell value: $p = 0.55$.

The obvious problem with the quadrat method is in fitting the lifestyle data to a grid structure. That structure is arbitrary and encounters the modifiable areal unit problem (Chapter 8). Changing either the zoning (by shifting the grid origin or by rotation) or the scale (by changing the cell length or shape) will produce different results. The estimates are also based on differing sample sizes per grid cell – some cells contain more points than others.

9.1.2 Thiessen polygons

Arguably, the geography of Figure 9.4 is less arbitrary than the grid-based one. It shows what are known as Thiessen, Voronoi or proximity polygons. These are constructed such that any location within the boundary of a polygon is closer to the single sampled point also within the polygon than it is to any other sampled point. For consistency, Figure 9.4 is shaded in the same way and with the same class values as Figure 9.3. In practice,

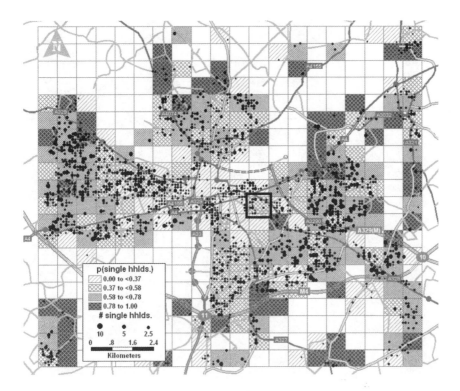

Figure 9.3 Grid aggregation of the lifestyle data, displaying the proportion of 'singles households' per grid cell

the range of possible values is constrained, as discussed previously, due to there being only one sample point per polygon from which to derive an estimate of the proportion of singles households. Although the sample size is now broadly consistent (insofar as $n = 1$ postal unit in all cases) it is also too small!

As it happens, Thiessen polygons are constructed by first undertaking Delaunay or Dietrich triangulation that connects sampled points to their nearest neighbours. The result is a tessellation of triangles covering the study region and is often referred to as a TIN (triangulated irregular network, see Chang, 2003, p. 281). An estimate of the attribute value at a non-sampled location can therefore be made by identifying the triangle containing the location and then using the three sample points at the corners of the triangle to derive the estimate within. The sample size is consistent and larger than previously but remains small ($n = 3$).

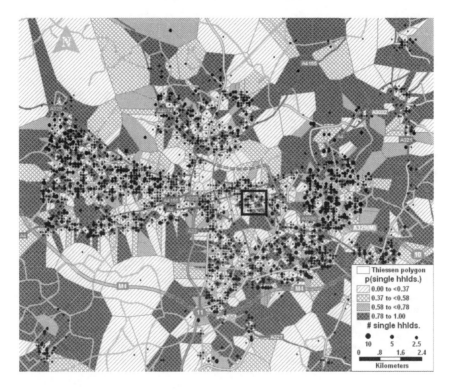

Figure 9.4 Polygonal aggregation of the lifestyle data, displaying the proportion of 'singles households' per Thiessen polygon

9.1.3 A concentric circle approach

An alternative way of estimating the proportion of singles households in and around the non-sampled location is to adopt one of the catchment-defining methods outlined in Chapter 5, Section 5.2.5. Figure 9.5 adopts the simplest method, drawing a circle around the estimation point and calculating the proportion of singles households within that buffer zone. In fact, Figure 9.5 shows the edges of a series of concentric circles, each centred at the same point but with different radii: the first has 500 m; the second 1000 m; and so forth, increasing at 500 m intervals to 2500 m. A problem, as with the earlier grid-based approach, is that the estimated value is modifiable, dependent in this example on the scale of analysis (the radius length). A short radius of 500 m provides an estimate of $p_{1\,(500\,\text{m})} = 0.47$. Increasing the radius to 1000 m produces $p_{2\,(1000\,\text{m})} = 0.36$ (a 23% decrease); whereas a radius of 2500 m produces a result similar to the first, at $p_{3\,(2500\,\text{m})} = 0.46$.

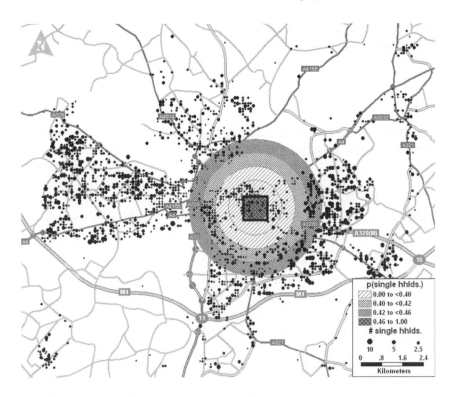

Figure 9.5 Concentric ring aggregation of the lifestyle data, displaying the proportion of 'singles households' per buffer zone

9.1.4 Inverse distance weighting

Instead of calculating the buffers as five overlapping circles, they could instead be generated as they seem to appear in Figure 9.5: as five separate and sequential rings, the second ring starting at the outer edge of the first 'ring'; the third ring starting at the outer edge of the second; and so forth. The midpoint distances, d_r, between the inner and outer edges of these rings from the central point outwards are: 250 m; 750 m; 1250 m; 1750 m and 2250 m. The corresponding estimates of the proportion, p_r, of singles households within each ring are: 0.47; 0.33; 0.44; 0.45 and 0.51, respectively.

It is possible to 'pool' the data for each ring and obtain an inverse distance weighted estimate of the proportion of singles households around the central point. In accordance with Tobler's first law of geography (Chapter 1, Section 1.5), the average will give increased weight to the ring (r) at least distance, d_r, from the centroid point and decreased weight to

255

the ring that is furthest away. Specifically, the weight we assign to each ring will be defined as:

$$w_r = \frac{(2750 - d_r)^c}{\Sigma_1^r(2750 - d_r)^c}$$ (9.1)

Equation (9.1) ensures that each ring contributes to the average but also that the weight assigned decreases with distance from the estimation point, reaching zero beyond the fifth ring (at $d = 2750$ m). The rate of decay is controlled by parameter c ($c > 0$ for inverse distance weighting). Setting $c = 2$ specifies an inverse squared decay and produces the following set of weights, from the centre outwards: 0.45; 0.29; 0.16; 0.07; 0.02. Given that the five weights sum to one, the pooled and inverse distance weighted average is calculated as:

$$
\begin{aligned}
\bar{p} = \sum_{r=1}^{5} w_r p_r \\
= (0.45 \times 0.47) + (0.29 \times 0.33) + (0.16 \times 0.44) \\
+ (0.07 \times 0.45) + (0.02 \times 0.51) \\
= 0.42
\end{aligned}
$$ (9.2)

This weighted average better reflects the attributes of the inner ring than it does each successive ring – by definition. Increasing the value of c places further importance on the near rings relative to the far ones. When $c = 3$ the weighted average equals 0.43 and when $c = 5$ the weighted average is 0.44. As c is increased, the weighted average is becoming increasingly like the average for the first ring alone ($p_1 = 0.47$).

Inverse distance weighting (IDW) is often incorporated in methods of spatial data manipulation such as population surface modelling (Bracken and Martin, 1989; Martin and Higgs, 1997; Martin, Tate and Langford, 2000) and in methods of spatial data analysis such as geographically weighted regression (Fotheringham, Brunsdon and Charlton, 2002). By doing so these methods recognize the existence of spatial autocorrelation (see Chapter 1, Section 1.5) and, by allowing the weights to decay to zero from a centre point outwards, avoid imposing a stark boundary between points that are either fully in an analytical area or fully out of it. Such boundaries are present in both the grid-based and Thiessen polygon methods (Figures 9.3 and 9.4) where there are sharp discontinuities at the edge of each area unit. However, IDW methods require consideration be given to the shape of the weighting kernel which is modifiable both in terms of the distance over which it operates and also, as we have seen, the importance given to near data points relative to those further away. In practice, the distance decay parameter has less effect on the analytical results than the maximum length/radius of the kernel – the effects of which can be considerable (see, for example, Harris and Chen, 2004).

9.2 Looking for 'hot spots'

For the purposes of this chapter, the phrase 'hot-spot analysis' is treated as shorthand for using spatial analytical techniques to look for and adjudge the significance of geographically clustered patterns of spatial autocorrelation in attribute values taken from lifestyle datasets. We could also describe this as point pattern analysis since we have been assigning point coordinates to georeference and map the lifestyle data. However, unlike a number of point pattern techniques that are only concerned with summarizing the pattern of the locations themselves (the geography of the sample), what we are interested in is the geographical patterns of the attribute values assigned to those locations.

9.2.1 A concentric circle method

A simple framework for hot-spot analysis is suggested by the circular method of interpolation considered in Section 9.1.3. But instead of generating a (single) circle of fixed radius around a non-sampled location, imagine that a circle was produced around each and every sample point (one circle around each point, *not* multiple circles as in Figure 9.5). Doing so would permit summary statistics to be calculated for points falling within each of these circles, for example the local proportion of singles household respondents. The local statistics could then be compared against each other and against 'global' values for the study region as a whole, permitting potentially unusual and localized occurrences to be identified.

This is the general approach adopted by one of us (Harris, 2004) to look for localized patterns of consumer behaviour present within the City and Country of Bristol, England. A slight modification is that in order to compensate for the geographically uneven distribution of survey responses it was not the circle radii that were fixed but the target number of survey respondents within each circle. This involves identifying the n_i nearest neighbours to any point (i) to give a total local sample of m respondents around and inclusive of i, holding m constant from point to point. Having identified the n_i nearest neighbours they are grouped together to form set I and a calculation made of the proportion of respondents possessing a certain attribute within the set (e.g. the proportion of singles household respondents).

Imagine that point j is one of those n_i nearest neighbours – a member of set I. In the same way that the closest points to i can be identified so too can the n_j nearest neighbours to j. Call this second group set J. Again, a calculation can be made of the proportion of 'single' respondents, this time within set J. Repeating the process, equivalent measures can be found for each of the n_i points around i. The result is n_i estimates of the proportion of singles household respondents within the vicinity of location i – one for each member of set I.

By comparing the estimates, measures of the local mean $(\bar{a}_i)$ and the local standard deviation (s_i) around that mean can be calculated. The global proportion (μ) of 'single' respondents across the entire study region (or 'universe') can separately be calculated. Combining all this information, a measure of the relative significance of the attribute's concentration – in this case 'singleness' – in and around i is derived as:

$$z_i = \frac{\bar{a}_i - \mu}{s_i/\sqrt{n_i}} \tag{9.3}$$

The index value, z_i, is highest in localities that have a higher than global average presence of singles household respondents *and* where that concentration is not a 'one off' but generally characteristic of all the n_i neighbours of i. Based on this index, Figure 9.6 suggests hot spots within the southern England town where singles households particularly are located – predominantly on the periphery of the settlement.

Equation (9.3) deliberately mimics the formula commonly employed to undertake a z test of the difference, along a normal distribution, between a sample mean $(\bar{x})$ and the hypothesized mean (μ) for the population from which the sample is drawn, i.e.

$$z = \frac{\bar{x} - \mu}{s/\sqrt{n}} \tag{9.4}$$

(cf. Rogerson, 2001, p. 44).

However, the results of Equation (9.3) must be interpreted differently from a standard z test because the normality assumptions do not apply. This means we cannot use standard statistical tables to conclude that a z value of magnitude greater than 2.0 has less than 5% likelihood of occurring 'by chance'. How, then, can we assign significance to the

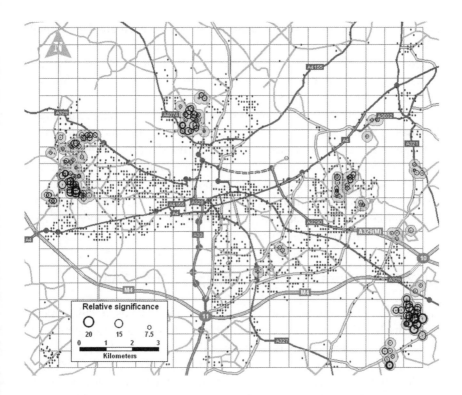

Figure 9.6 Results of 'hot-spot analysis' detecting significant clusters of 'singles households'

z_i values resulting from Equation (9.3)? At what threshold should we become interested in the result?

Looking back at Figure 9.2 suggests the answer. Recall that this map has the correct geography of the lifestyle sample but the attribute values at each location randomly were redistributed around the scene. Running the hot-spot analysis on this quasi-random dataset it is found that approximately 5% of the sample are given z_i values less than or equal to -3, or greater than or equal to $+7$. These results imply that given the geography of sampled locations and given the set of records recording the number of singles households and the number of respondents, z_i values above $+7$ would occur, by chance, 2.5% of the time. In other words, we can have 95% confidence that $z_i \leq -3$ or $z_i \geq +7$ are not 'false positives'. For our sample of lifestyle data the $+7$ threshold might then be treated as comparable to the $+2$ cut-off of the conventional z test.

However, the $+7$ threshold is case specific and, at this stage, not reliable. Remember that it derives from a random redistribution of the sample attributes and could, therefore, be freakish. We should repeat the process of testing for random redistributions of the attribute data perhaps 100, 500 or 1000 times until a clear idea is formed of what threshold values genuinely are unusual. Such a process of repeat testing to determine benchmark values is known as Monte Carlo simulation (see O'Sullivan and Unwin, 2003, pp. 104–8).

As it turns out, a threshold of +7 is not unreasonable and therefore has been applied in Figure 9.6, above. Using the techniques described in Chapter 5 to cross-reference the neighbourhood profiles of the postcode sectors in which these significant clusters are found against the profiles for all sectors returning a lifestyle respondent, singles households are found to have a higher than expected presence in 'Symbols of Success' neighbourhoods (index value of 187), 'Suburban Comfort' (index of 135) and 'Grey Perspectives' (111).

9.2.2 The geographical analysis machine (GAM)

A definition of spatial analysis is 'a set of methods whose results change when the location of the objects being analysed change' (Longley *et al.*, 2001, p. 278). Another way of looking at this is in terms of the modifiable areal unit problem – when the scale or zoning of the analysis changes so too does the result. Accordingly, in the hot-spot analysis outlined above, if the circular buffer drawn around each point were repositioned or resized then different z_i values would be obtained. A weakness of the method is that there is only one circle drawn around any one point and its definition is essentially arbitrary. A better method might try out differently sized circles, retaining those that are significant to the local context and rejecting those that are not.

Such a process of repeat testing is at the heart of the 'geographical analysis machine' (GAM), an analytical procedure developed to test whether incidences of childhood leukaemia were clustered around British Nuclear Fuels' Sellafield site. They were but a stronger cancer cluster was found in Gateshead (Openshaw *et al.*, 1987, 1988). The GAM algorithm is described by Openshaw (1998, p. 97). It is paraphrased as follows:

- Read in (*x*, *y*) data for population at risk and an attribute of interest from a GIS.
- Identify: the minimum enclosing rectangle (MER) containing the data; the starting circle radius; and the degree to which neighbouring circles will overlap.
- Generate a regular, raster grid across the MER with cell length defined by the amount of overlap between circles.

- For each grid intersection (cell corner) generate a circle of radius r.
- Retrieve two counts for the circle: the local population at risk and the local attribute count.
- Apply some 'significance' test procedure and keep the result if it is significant.
- Repeat for all remaining grid intersections.
- Increase the cell radius, regenerate the grid and repeat until either there are no further significant circles or a maximum search radius is reached.
- Draw a smoothed density surface showing how the number of significant circles varies across the study region.

Plate 4 maps the result of using GAM to search for unusual clusters of the lifestyle survey respondents classifying themselves as 'single'. Also shown are the results of the previous, single circle per point method. Reassuringly, there is much accord: the majority of localities assigned significance in one test is also assigned significance in the other. There are differences, too. These are not entirely unexpected since the methods do not assign significance in the same way. The current incarnation of GAM offers a variety of methods for detecting clusters in spatial and temporal datasets. It can be downloaded from the Centre for Computational Geography at the University of Leeds: www.ccg.leeds.ac.uk.

GAM is an example of a geographical data-mining tool used, like neighbourhood classifications, for knowledge discovery or, more formally, for inductive reasoning. Induction is to draw a general rule or conclusion from particular facts or examples – in this case, a particular dataset. It contrasts with deductive reasoning for which the general rule or conclusion is first decided and then tested by appropriate means such as data collection and analysis, to further a conclusion. In the context of geographical information science, induction could be viewed as data-led, deduction as theory-led. However, as the term 'data-led' is sometimes intended to be pejorative it may better be avoided. In reality, within the 'messy' and subjective world of research and analysis the division between induction and deduction is often blurred.

9.3 From revelation to explanation

As well as describing GAM, Openshaw (1998) also introduces a geographical explanations machine to automate the search for localized spatial associations between detected clusters and predictor variables that may help to explain why the clusters have formed. A caveat is important here to avoid misunderstanding of this and other associative methods: finding spatial associations can *help* to explain why particular phenomena or events are found to have greater presence in some places than others but does not, in itself, prove cause and effect. Sales of ice cream and sales of barbeque fuel are correlated but it of course does not follow that eating a vanilla cone causes a subsequent meal to be roasted outdoors!

A method of detecting localized spatial associations, which is becoming more prevalent, is geographically weighted regression (GWR, Fotheringham, Brunsdon and Charlton, 2000). This brings together the ideas of moving search windows and inverse distance weighting to quantify how the measured associations between variables vary spatially across a study region. It avoids the 'one size fits all' modelling of traditional approaches to allow important local relationships to be measured and mapped in a GIS. It is an explicitly geographical technique in that it does not attempt to 'average away' spatial variations and because it does not make the false assumption of much linear regression modelling of geographical datasets that any residual error is random over space. To the contrary, patterns of geographical dependence are expected to be present in the dataset and it is these the modelling procedure aims to reveal and help explain. Software for GWR is available via www.may.ie/ncg/gwr. An application of using GWR to model the spatially varying correlates of urban deprivation is provided by Longley and Tobon (2004).

Another, method for revealing geographical variations in the associations of variables is multilevel modelling (Snijders and Bosker, 1999; Hox, 2002; Goldstein, 2003). Although not exclusively a geographical technique, many of the 'units' for which geographical data are collected do have a hierarchical structure – i.e. multiple levels – such as pupils within schools within local authorities, or households within census output areas within electoral wards within governmental regions, etc.

Multilevel modelling can be likened (but is not limited) to fitting a regression line, $y' = \beta_0 + \beta_1 x$ to a set of data but allowing the slope (β_1) and/or y-intercept (β_0) coefficients of that line to vary from place to place at the various levels of the hierarchy. This permits the significance of the variance at each level to be determined and can be used to examine Weiss' (2000, p. 8) assertion that there is a 'basic clustering concept, that people in the same neighborhoods tend to behave (or at least consumer) the same way.' For example, a three-level model of household lifestyle data linked to a classification of neighbourhood type and the region within which the neighbourhood is located could examine the relative significance of household, neighbourhood and regional variations with regards to a particular consumer attribute. Alternatively, formulae for measuring deprivation such as the breadline one of Chapter 2, Equation (2.2) could be remodelled to take into account neighbourhood and regional differences in the correlates of deprivation.

A considerable advantage of multilevel modelling is that it is an integrated methodology in that the various levels of the hierarchy are coherently linked together. One benefit of this is parsimony in that the variations in the parameter estimates obtained for the various parts of the hierarchy are incorporated and thus comparable within a single model. A second is of obtaining more robust parameter estimates both by pooling data and by not ignoring the effects of spatial autocorrelation within the units of analysis (effects which otherwise lead to the significance of the parameter values to be overestimated). However, use of the method assumes that a hierarchical structure (not necessarily geographical) is present and that its units and levels can be defined. While pupils belong to schools that have clearly defined boundaries, geographical units for analysing consumer behaviour are more indeterminate. Multilevel modelling software is available from http://multilevel.ioe.ac.uk.

9.4 Data-handling issues

9.4.1 Data uncertainty

A potential problem with lifestyle datasets, mentioned in the case study and also considered by Longley *et al.* (2001), is the 'uncertainty' of the data. As those authors note, any digital representation of the real world is imperfect and uncertainty is introduced through the ways in which the world is conceptualized, measured, analysed and represented. Uncertainty is an umbrella term covering issues of error, inaccuracy, ambiguity and vagueness.

With lifestyle datasets the focus is on uncertainty in the measurement and representation of people and places. However, these notions of uncertainty cannot be separated from the applications to which the data are put. Provided that the survey returns are up to date and truthful we can have a large degree of certainty that lifestyle data provide useful and accurate information about the people who complete lifestyle surveys. Of course, the information provides only a snapshot or caricature of the 'complete person', but lifestyle surveys are often extensive and the information that is gleaned from them especially is relevant to direct marketing.

Greater uncertainty arises when the data are extrapolated to profile people and places that have not returned lifestyle information. This is because the data provide only a sample of the broader populace and caution is needed when remodelling the data in ways that assume they are representative of a non-sampled group. As it happens, if we profile the neighbourhood characteristics of the lifestyle sample analysed in the preceding sections against what we would expect to find within the local authority concerned they are extremely consistent: the Pearson correlation of the observed and expected distributions by neighbourhood group is +0.99 (99%). Admittedly, the most affluent neighbourhoods (Affluent Achievers, Happy Families and Suburban Comfort) are relatively oversampled with index values of 112, 113 and 117, respectively, and the Welfare Borderline group are the most undersampled with an index value of 72. However, strict notions of a 'digital divide' are, in this case, not entirely tenable: relatively, the most oversampled group turns out to be the Municipal Dependency group, with an index of 123.

Nevertheless, the 'representativeness' of lifestyle data cannot always be assumed, because of the way the data are collected. Lifestyle samples are unusual in that those who are enumerated are partly self-selecting in the sample design (some people opt out of receiving unsolicited mailings), partly self-selecting as respondents (having received a survey they are under no obligation to return it), in part selected by the data company (few surveys attempt a blanket coverage of the population but are targeted to particular households or attached to particular types of purchases) and in part determined by the respondents' life circumstances (certain groups are more able or willing to return a survey than others; some are more attracted to incentives to do so, such as shopping vouchers or a prize draw; some are more likely to buy the products to which extended warrantee questionnaires are attached). The sample therefore is determined by a conflation of random and systematic events.

One way to ascertain how representative the data are is test them against some suitable benchmark. Above and in Chapter 5, Sections 5.2.1 and 5.2.4, the benchmark was provided by comparing the observed and expected mix of neighbourhoods. This is a useful and quick method of analysis but, in its crudest form, will only 'correct' the data by assigning the same weight to all respondents within a neighbourhood group, therefore discounting any measurable differences in the lifestyle attributes of the group that may be important in determining response. A more sophisticated method is outlined as follows.

Imagine that a recent national census had revealed the number of households in zone A to be 1000. For the same area, a lifestyle survey had enumerated 100 households. To tally the numbers, each survey responding to household needs to count for 10 in the census (i.e. $10 \times 100 = 1000$).

Assume that of the 1000 households, 600 are recorded in the census to be possessing at least one car. The lifestyle survey has 70 respondents owning a car, which is an overestimate relative to the census, since $70 \times 10 = 700$ and $700 > 600$. If we trim the weight assigned to each car-owning household in the survey to 8.57 then the correct census count is obtained: $8.57 \times 70 \approx 600$. However, the total number of households now falls short: (car-owning households + no-car households) $= (8.57 \times 70 + 10 \times 30) \approx 900$ and $900 < 1000$. We cannot change the weight assigned to the car-owning group if their weighted total is to remain at 600, so the only option left is to raise the weight assigned to no-car households, to 13.3. This gives the correct total: $(8.57 \times 70 + 13.3 \times 30) = 1000$. We now have two weights that we can assign to lifestyle respondents in zone A: the first for those who own cars (8.57); the second for those who do not (13.3).

Further assume that the census counts 400 households living in terraced properties in area A. The lifestyle survey has 25 households in terraced properties and, querying the database, it is found that 20 of the 25 own cars (five, therefore, do not). Based on the current weightings, the lifestyle survey has a weighted total of $(8.57 \times 20) + (13.3 \times 5) \approx 238$ households living in terraced properties – a shortfall of 162. Assigning 20/25ths of the shortfall to the car-owning, terraced households, and 5/25ths to the non-car-owning, terraced households produces weights of 15.1 and 19.8, respectively ($15.1 \times 20 + 19.8 \times 5 \approx 400$).

Unfortunately, the weighted count of car-owning households is no longer correct. It is known that 20 of the 70 survey respondents who own a car live in terraced properties and the weight assigned to this group is 15.1. For the remaining 50 who do not live in terraced properties the only weight we have available for them is the one first derived for the car-owning group as a whole, i.e. 8.57. Putting these together gives a weighted count of $(15.1 \times 20) + (8.57 \times 50) \approx 731$ car-owning households, which is a surplus of 31 over the corresponding census count. The weight assigned to car owners living in terraced properties cannot be changed if the weighted count of terraced households is to stay at 400; instead, the weight assigned to car-owning households not living in terraced properties is lowered by the required amount, to 7.95.

The weighted counts of all terraced and all car-owning households are now consistent for the lifestyle sample and the census, in area A, but the all household counts are not! That total is found as the weighted sum of: car-owning, terraced households; non-car-owning, terraced households; car-owning, non-terraced households; and non-car-owning, non-terraced households. Of these four classes of household, weights have been adjusted for all but the non-car, non-terraced group, for which the closest match is the weight originally found for the non-car-owning group, i.e. 13.3. The present weighted count of households from the lifestyle data is $(15.1 \times 20) + (19.8 \times 5) + (7.95 \times 50) + (13.3 \times 25) \approx 1131$. The only weight that can be adjusted without affecting either the weighted count of car-owning households or the weighted count of terraced households is that assigned to the non-car-owning, non-terraced households group. Trimming it from 13.3 to 8.06 gives the correct census count of 1000 households and we now have four classes of adjusted weight that we can assign to different types of lifestyle respondent in zone A.

Introducing a third comparative census variable, such as the number of households in privately rented dwellings, will produce eight weighting classes: car-owning, terraced households in rented dwelling; non-car-owning, terraced households in rented dwelling; car-owning, non-terraced households in rented dwelling; non-car-owning, non-terraced households in rented dwelling; car-owning, terraced house-holds, not in rented dwelling; non-car-owning, terraced households, not in rented dwelling; car-owning, non-terraced households, not in rented dwelling; and non-car-owning, non-terraced households, not in rented dwelling. As previously, weights are assigned and adjusted for classes sequentially.

First, the lifestyle database is queried to find the number of sample households living in area A in rented dwellings and, of those, the numbers living in each of the four classes for which weights previously have been derived, i.e. the number living in a terraced property and owning car; the number living in a terraced property but not owning a car; the number not living in terraced property but owning a car, and the number neither living in a terraced property nor owning a car. From this information the total weighted count of households living in rented accommodation in area A can be found, compared against the known census quantity and any surplus or deficit in the lifestyle estimate corrected by trimming or increasing the weights.

Next, the weighted count of terraced properties is recalculated from the lifestyles data using: the new weights found for terraced house-holds, car owning and renting, and for terraced households, not car own-ing but renting; and, the existing weights for terraced households, car owning but not renting, and terraced households, neither car owning nor renting. Those latter two weights (only) can be adjusted to compensate for any surplus or deficit against the census count of terraced properties with-out also affecting the weighted count of rented properties.

Attention now turns to recalculating and comparing the weighted count of car-owning households with the census value. The count is determined from the lifestyle sample as the weighted sum of: car-owning, terraced and renting households; car-owning, terraced but not renting households; car-owning, not terraced but renting households; and, car-owning but neither terraced nor renting households. Only the weight attached to the last of these categories can be adjusted without affecting the weighted counts of either renting or terraced households.

Finally, the weighted count of households is recalculated and adjusted to fit the census value. The lifestyle count is obtained by the weighted sum across all eight classes: car, terraced, rented; no car, terraced, rented; car, not terraced, rented; no car, not terraced, rented; car, terraced, not rented; no car, terraced, not rented; car, not terraced, not rented; and, no car, not terraced, not rented. Only the weight assigned to the last of these classes can be adjusted without altering the weighted counts of all renting, all terraced and all car-owning households.

In principle the process of introducing new comparative values and using them to refine the weighting of the lifestyle data could continue until there are no more variables left to compare. However, the method described is dependent upon being able to make the adjustment that aligns the weighted lifestyle count with the corresponding census value which, in turn, requires that the 'all not' class (not car owning, not terraced, not renting, etc.) is always populated. In practice the weighting may need to offer a best but not perfect fit between the lifestyle and baseline values.

Having obtained weights for the lifestyle sample in area A the process is repeated for areas B, C, D and so forth. In this way, weights can be fitted to the lifestyle sample that vary both spatially and with respect to the attributes of the respondents. Whether the weights should be operationalized within an analytical context depends on how representative (or otherwise) the raw data are of the baseline population and, more importantly, how representative the baseline data themselves are of the underlying population. At the time of writing it may be appropriate to fit the results of a recent consumer survey to what is revealed by the national census but as the time from the last census increases so our confidence in it as a suitable baseline will diminish. Finally, it is important to remember that the higher the weight assigned to a lifestyle respondent, the more under-represented that particular class of person or household is in the dataset. In other words, the higher the weight, the more unusual the response, which leads to uncertainty about quite how typical of the class that respondent's information actually is.

9.4.2 Data privacy and protection

Related to issues of uncertainty are questions of personal data protection. Personal data are information about individuals which is held on computer or is on paper and sorted by reference to individuals. Within the European Union (EU), the protection of individuals with regard to the processing of personal data is determined by Parliament and Council Directive 95/46/EC (dated 24 October 1995) and by Regulation Number 45/2001 of the European Parliament and of the Council (dated 18 December 2001) (see www.europa.eu.int). The Directive and Regulation together harmonize national laws across EU member states on the processing of personal data, with the intention of protecting the rights and freedoms of the persons concerned, in particular their right to privacy. They apply to the processing of personal data wholly or partly by automatic means and to the processing other than by automatic means of personal data which form part of a filing system or are intended to form part of a filing system.

Within the UK, the Directive is enforced under the 1998 Data Protection Act, making eight principles of handling personal data statutory. They state that data must be:

- fairly and lawfully processed;
- processed for limited purposes;
- adequate, relevant and not excessive;
- accurate;
- not kept for longer than is necessary;
- processed in line with people's rights;
- secure; and
- not transferred to countries without adequate protection.

Under the Data Protection Act, individuals are entitled to find out from companies the information held about them. If necessary, the individual may apply to a court for the rectification, blocking, erasure or destruction of inaccurate personal data and any other personal data in respect of which s/he is the data subject (see: www.informationcommissioner.gov.uk).

The EU's Directive prohibits the transfer of personal data to non-EU nations not meeting the standard set down for privacy protection. Initially this included the USA which does not rely on comprehensive legislation to ensure data protection but, instead, a mix of legislation, regulation and self-regulation. In order to bridge between these different approaches, the US Department of Commerce and the European Commission have developed a 'safe harbour' framework, under which transborder dataflow is permitted. Under the safe harbour scheme, organizations must adhere to seven principles:

- Notice – organizations must notify individuals about: the purposes for which they collect and use information about them; how they can contact the organization with any enquiries or complaints; the types of third parties to which information is disclosed; and the choices and means the organization offers for limiting how the information is used and disclosed.
- Opt out/opt in choices – organizations must give individuals the opportunity to opt out of the disclosure of their personal information to a third party or for a use incompatible with the purpose for which the data originally were collected. For sensitive information, an explicit opt in is required for information disclosure or transfer to a third party.
- Onward transfer to third parties – in addition to the notice and choice principles, if an organization wishes to transfer information to a third party acting as an agent, it may do so if that third party subscribes to the safe harbour principles, is subject to the EU Directive or enters into a written agreement of providing at least the same level of privacy protection as is required by the safe harbour framework.
- Access – individuals must have access to the personal information held about them and be able to correct, amend, or delete inaccurate information, except where the burden or expense of providing access would be disproportionate to the risks to the individual's privacy or where the rights of persons other than the individual would be violated.
- Security – organizations must take reasonable precautions to protect personal information from loss, misuse and unauthorized access, disclosure, alteration and destruction.
- Data integrity – personal information must be relevant to the purposes for which it is used. Reasonable steps should be taken to ensure data are reliable for the intended use, are accurate, complete and current.
- Enforcement – to ensure compliance with the principles, there must be methods of sanction, and available and affordable independent

mechanisms to: investigate individual complaints and disputes; verify the adherence of organizations to the safe harbour framework; and remedy problems arising out of a failure to comply.

(see: www.export.gov/safeharbor)

In the UK, one consequence of a rising interest in personal data protection has been the withdrawal of the full electoral register as a means to marketers of gathering information about the demographic composition of households, and about the names and residential addresses of their members. A consequence is that the full register is also no longer available for use as a sampling framework for lifestyle surveys or for modelling the characteristics of individuals and/or households for whom lifestyle information is not otherwise available.

The withdrawal of the electoral register was brought about by the '*Robertson* case' of November 2001. Prior to this case and its legal ruling, electoral registration officers were required to disclose the full register to anyone paying the appropriate fee. The register is a public document and has been long available in local centres such as public libraries for members of the public to examine and check. In principle, it was always possible for anyone to go from centre to centre and make a complete copy of the entire paper documents. In reality, it was only with the ability to collate and store the entire roll digitally that it entered into widespread use in many areas of marketing and surveying.

UK residents are required annually to supply personal information to electoral registration officers on penalty of a criminal offence. However, officers had no right to limit the disclosure of that information to third parties. In 2001, Mr Robertson challenged the refusal of Wakefield Council to not supply his personal information to commercial organizations. The High Court ruled that the refusal was in breach of both the European Data Protection Directive and the European Convention on Human Rights (since a consequence of Mr Robertson not completing an electoral return was that he would not be entitled to vote).

The aftermath of the court's decision was that local authorities stopped supplying the electoral register to commercial organizations. Subsequently, 'two' electoral registers have been developed, permitting residents to opt out/opt in to supplying their details to commercial companies (other than credit-referencing agencies or those involved in preventing fraud, which have access to the full register). As we stated in Chapter 3, in December 2002 the opt-out rate was just above 20% but has risen to 26% from the December 2003 register. The rate varies geographically with different authorities operating various opt-out/opt-in schemes. In part a response to this situation, some companies have encouraged 'data exchanges' to share consumer records (see, for example, www.eurodirect.co.uk).

9.5 Conclusion

In this chapter our attention has been on lifestyle datasets. These are of immediate use in direct marketing, targeting those consumers who supply information about themselves. The focus here, however, has been broader, looking into how the data might be used to produce what we have referred to as geolifestyle classifications, employing bottom-up techniques of analysis. Such techniques are drawn from the broad field of geographical information science, including GIS, geocomputation and spatial analysis to search for and, in the more sophisticated approaches, help explain spatial patterns in the attributes of the lifestyle data. They include grid-based aggregations, Thiessen polygons, buffering, inverse distance weighting, methods of hot-spot detection and geographically weighted regression. Multilevel modelling is not explicitly a geographical approach, although geographical units of enquiry often form a hierarchy that can be modelled using multilevel methods.

Issues of data uncertainty and of data protection are important to consider when handling and analysing lifestyle datasets. Particular care is required when extrapolating lifestyle information to model the characteristics of those for whom no or little information is held. Within the EU, for instance, personal data storage is required to be (among other stipulations) adequate, relevant and accurate. Lifestyle surveys and databases offer extensive but nonetheless partial coverage of populations and so any remodelling and analysis of the data should proceed carefully. Notwithstanding this important caveat, one of us has presented a number of case studies of the potential utility of such datasets in helping to understand the scales and localities of urban deprivation (Harris and Longley, 2004), complementing more traditional geodemographic approaches (Longley and Harris, 1999).

Summary

- Lifestyle data is an umbrella term that covers information collected about consumer choices and behaviours.
- Such data are of importance to one-to-one and direct marketing but can also be modelled to create geolifestyle classifications.

- Techniques of spatial analysis can be used to look for spatial patterns in the attributes of lifestyle datasets.
- Methods that employ inverse distance weighting reflect Tobler's first law of geography – everything is related to everything else but near things are more related than far things.
- Determining the significance of detected 'hot spots' often requires methods of simulation when the assumptions of standard statistical tests are violated.
- The geographical analysis machine (GAM) is an exploratory and geographical data-mining tool that detects spatial clusters in datasets.
- Methods such as geographically weighted regression avoid the 'one size fits all' modelling of traditional approaches to allow important local relationships to be measured and mapped in a GIS.
- Modelling, analysing and extracting information from lifestyle datasets require attention to issues of data uncertainty, data protection regulations and good practice.

Further Reading

- Bailey, T. and Gatrell, A. (1995) *Interactive Spatial Data Analysis*, Prentice-Hall, Harlow, Essex.
- Longley, P., Goodchild, M., Maguire, D. and Rhind, D. (2001 and 2005) *Geographic Information Systems and Science*, Wiley, Chichester.
- Monmonier, M. (2002) *Spying with Maps: Surveillance Technologies and the Future of Privacy*, University of Chicago Press, Chicago.
- O'Sullivan, D. and Unwin, D. (2003) *Geographic Information Analysis*, Wiley, New York.

10
Postscript: There are three Is in geodemographics!

It may be a while since you read it (assuming you did!) but cast your mind back to Chapter 2 and the origins of geodemographics. You may recall that three factors contributed to the development and growth of the field. First, an interest in understanding the social and spatial structure of cities and their neighbourhoods. Second, the increased availability of census statistics that could be used to characterize different types of locality. Third, new methods of statistical analysis by which to make sense of the new data. Together, these factors are the three Is of geodemographics – inclination, information and innovation.

These three Is not only gave birth to a whole new industry but also have nurtured its growth and evolution. What is more, they are mutually reinforcing. As we have seen, renewed interested in neighbourhood regeneration within the UK has given rise to new ways of understanding and quantifying area deprivation that have incorporated new sources of neighbourhood data that, in turn, have been made available to other geodemographic users through a specially developed internet portal. This and other portals are making the exchange and purchase of geodemographic data faster and easier than ever before, and purchasers have more flexibility to buy exactly what they need for their applications. Software such as GIS and more marketing-focused geodemographic information systems make analysis, understanding and presentation of such data easier, generating new knowledge, ideas and perspectives. In the fields of spatial statistics and mathematics more generally, new analytical approaches are being

Geodemographics, GIS and Neighbourhood Targeting   Richard Harris, Peter Sleight and Richard Webber
© 2005 John Wiley & Sons, Ltd   ISBNs: 0-470-86413-3 (HB); 0-470-86414-1 (PB)

developed to help explore and explain the patterns of spatial association that are revealed in geographically and temporally precise sources of data. Developments in geosimulation offer new insights into the dynamics of urban systems.

We live in an information society and so new sources of data keep on being generated, some of which are sector specific, while others have more general application. Applying clustering techniques to the more specific data allows 'niche' products to be developed, with an expectation that they will necessarily perform better within that sector than the more general classifications also available. There is evidence that in terms of the more customized products that are appearing the geodemographic market is fragmenting. However, a word of caution is appropriate here. The suggestion that 'for any commercial (or indeed non-commercial) application, a general purpose [geodemographic] system is likely to be inferior to one designed specifically for the purpose' (Flowerdew, 1991, p. 8) is intuitive but are things actually so clearly cut?

A practical consideration is the role of the more general-purpose classifications as the 'lingua franca' that allows analyses to be compared – the common currency argument raised in Chapter 8. The value of this (the strength of the geodemographic pound, as it were) should not be under-estimated; however, there is also a more 'technical' explanation why the more customized products are not *necessarily* the better. It is because they risk being overly calibrated on a (narrow) set of data that actually describe past events. It may not be the more specific past events that better explain future behaviour (the fact that the sun shone yesterday is no guarantee it will shine today, as weather forecasters know well) but the more enduring general factors – climate cycles in the case of weather, the broader socio-economic and demographic landscape in the case of consumption.

The critical issue is one of generality versus specificity. Barnsley, Moller-Jensen and Barr (2001) consider this in the context of urban remote sensing – a field that is not so far removed from geodemographics as it may seem, but actually employs some similar and transferable clustering tech-niques (for example, to segment a remotely sensed image into identifiable land-use classes). They see advantages of a generic model in the power of constructing a relatively simple but understandable representation of a large number of objects (e.g. neighbourhoods) and in the portability of that general model from one location to another, permitting comparison between places. The general model will not usually offer the same level of detail that might be expected from a more focused study of a particular place at a particular time but the problem with the more specific model is that, being attuned to a par-ticular place and time, it is not well transferred to other times and settings.

It will be interesting to see whether the future is one where geo-demographic vendors prefer to promote and develop the more general classifications, or whether more specific classifications will emerge from the consumer, financial, retail, education, b2b (business to business), telecoms, e-commerce and other markets. We have already seen that techniques such as K-means clustering are not limited to neighbourhood data but can be used to classify individuals and households directly. Products linking types of neighbourhood globally are also now available. Which company will be the first to attempt a global classification of all 6.4 billion of the world's population?!

It is tempting to conclude this book euphorically, envisioning a future for geodemographics that is limited only by the creativity and imaginations of classifications builders and the data available to them, or by the users' abilities to put the products and technologies to use in their businesses and corporations. Such a future may become real but, as portrayed, risks eschewing legitimate public concerns about the sorts of data that are held about them, the ways they are represented by those data and the control they can have over the way organizations use that infor-mation to manage their customers' relationships.

There is an uneasy tension of the era whereby consumers want (or, perhaps, are assumed to want) to be treated in a more individual, per-sonalized and less standardized manner by the suppliers of businesses and services. Yet, they often also seem to be unhappy with the use, by large organizations, of the types of data that would make this possible. A prob-lem is that many organizations – most likely for logistical reasons – are not especially open about the data they hold on people and are also not flexible about letting the consumer change it. But, why not let them? After all, it is the consumer's digital persona that is constructed, so might they be entitled to construct it in ways that they want? It makes sense from a marketing perspective: since it is how people see themselves and not how they are that usually is of prime interest to marketers (internet-based?) systems that allow consumers to volunteer such information are likely to become increasingly important and successful in the future (Hagel and Singer, 1999).

Current frameworks for data protection vary internationally and even within a single nation regulations are not especially consistent across 'different' types of data – the provisions made for census confidentiality and other governmental surveys, for example, are usually more rigid than for commercial surveying and data collection, and while there appears to be no (nontechnological) limits to how much 'snooping' is permitted by remote-sensing equipment, recent legislation is intended to leave email inboxes free from 'spam' (it does, however, remain available in the tinned meat section of many UK and US supermarkets – see www.spam.com for details).

Despite these inconsistencies, the general trend is one of 'clamping down' on the storage and transfer of personal data held on identifiable individuals (as opposed to aggregate, public datasets), and of limiting the types of inferences that may be made about individuals, especially without their consent. While such restrictions may prove awkward to some marketing operations in the short term, they may prove more beneficial over the longer run. First it may help to clarify which consumers want to receive certain types of mailings and information, and what is their preferred media to do so, thereby helping to avoid the resentment of consumers who are targeted by junk mailings they really did not want to receive. Second, it may have the unintended effect of reminding us that the world is not simply made up of bits of data that are waiting to be collected, stored and analysed. Numeric methods can tell a little of the story but it will likely be partial and incomplete. If we really want to understand consumers, neighbourhoods, motivations and lifestyles we need to move out of the computer world and experience the real one more deeply. Quantitative methods are not an alternative to qualitative ones. The best research employs both.

From the perspective of this book an interesting consequence of any restriction to the flow of lifestyle and other personal data will be to further galvanize the use of neighbourhood-based inference to try and fill 'holes' in databases. Weiss (2000, p. 8) reveals that he 'takes some comfort in the benign nature of the clusters, in the fact that they're designed to explain patterns of group behaviour without the need to delve into individual households.' Again, there is need for caution. The clusters may be benign but the applications may not be, as in the alleged but illegal incidences of 'red-lining' certain places from the provision of loans or insurance cover. There is an interesting debate to be had on whether it is better to base policy and decision making on precise and accurate data that can identify actual individuals/households, or whether less precision and accuracy is acceptable if it preserves anonymity, albeit at the risk of generalizing ('stereotyping') the actual characteristics of individuals. Discussion of a possible 'digital divide', about who data is collected about, why, how the data are used to frame public- and private-sector decision making and on how those decisions impact on the lives and well-being of both those represented and those excluded from the datasets is also important here. Also relevant are concerns about how initially separate datasets are being joined together on the basis of common georeferences such as name and residential address to create possibly far-reaching governmental and commercial profiles of the populace, and the ways such profiles can be linked to 'tracking' technologies such as GPS, Remote Sensing, CCTV or RFID (Radio Frequency Identification) chips.

Finally, as we reflect on the contents of this book we are aware that our abiding focus on geodemographics offers cohesion but may also have raised an unintended and inaccurate impression that it is a 'standalone' tool, competing with others to guide decision making. This is rarely, if ever, the case. Increasing integration of technologies and the increased ease of interoperability are likely to bring various decision-making tools together, especially where these have geography in common. Perhaps we will see even closer links between GIS, geodemographics, the visualization of geographic information and spatial analysis. Statistical packages like S-Plus (www.insightful.com) offer integration with GIS, while an open-source programming language such as R provides a wide variety of statistical and graphical techniques (www.r-project.org).

Whatever the future, we are confident that the three Is of geodemographics will keep vendors, users and the contributors to this book extremely busy for many years to come. And, what of us three? Do we envision our interest in geodemographics will endure? Our answers are obvious: 'aye!', 'aye!', 'aye!'

References

Anderson, B., Beinart, S., Farrington, D., Longman, J., Sturgis, P. and Utting, D. (2001) *Risk and Protective Factors Associated with Youth Crime and Effective Interventions to Prevent It* (Youth Justice Board Research Note 5), Youth Justice Board, London.

Baker, K., Bermingham, J. and McDonald, C. (1979) The utility to market research of the classification of residential neighbourhoods, *Proceedings of the Market Research Society Conference*, Brighton, pp. 253–71.

Barnsley, M., Moller-Jensen, L. and Barr, S. (2001) Inferring urban land use by spatial and structural pattern recognition. In *Remote Sensing and Urban Analysis* (eds, Donnay, J.-P., Barnsley, M. and Longley, P.), Taylor & Francis, London, pp. 115–44.

Batey, P. and Brown, P. (1995) From human ecology to customer targeting: the evolution of geodemographics. In *GIS for Business and Service Planning* (eds, Longley, P. and Clarke, G.), GeoInformation International, Cambridge, pp. 77–103.

Batty, M. (2003) Agent-based pedestrian modelling. In *Advanced Spatial Analysis: the CASA Book of GIS* (eds, Longley, P. and Batty, M.), ESRI Press, Redlands, CA, pp. 81–128.

Batty, M. and Longley, P. (1994) *Fractal Cities*, Academic Press, London.

Baudot, Y. (2001) Geographical analysis of the population of fast-growing cities in the third world. In *Remote Sensing and Urban Analysis* (eds, Donnay, J.-P., Barnsley, M. and Longley, P.), Taylor & Francis, London, pp. 225–41.

Benenson, I. and Torrens, P. (2004) *Geosimulation: Automata-based Modeling of Urban Phenomena*, John Wiley & Sons, Inc, New York.

Berry, B. and Horton, F. (1970) *Geographical Perspectives on Urban Systems*, Prentice-Hall, New Jersey.

Berry, M. and Linoff, G. (1997) *Data Mining Techniques for Marketing, Sales, and Customer Support*, John Wiley & Sons, Inc, New York.

Besussi, E. and Chin, N. (2003) Identifying and measuring urban sprawl. In *Advanced Spatial Analysis: the CASA Book of GIS* (eds, Longley, P. and Batty, M.), ESRI Press, New York, pp. 109–28.

Geodemographics, GIS and Neighbourhood Targeting   Richard Harris, Peter Sleight and Richard Webber
© 2005 John Wiley & Sons, Ltd   ISBNs: 0-470-86413-3 (HB); 0-470-86414-1 (PB)

References

Birkin, M. (1995) Customer targeting, geodemographics and lifestyle approaches. In *GIS for Business and Service Planning* (eds, Longley, P. and Clarke, G.), GeoInformation International, Cambridge, pp. 104–49.

Birkin, M., Clarke, G. and Clarke, M. (2002) *Retail Geography and Intelligent Network Planning*, John Wiley & Sons, Ltd, Chichester.

Birkin, M., Clarke, G., Clarke, M. and Wilson, A. (1996) *Intelligent GIS: Location Decisions and Strategic Planning*, GeoInformation International, Cambridge.

Blake, M. and Openshaw, S. (1995) *Selecting Variables for Small-Area Classifications of 1991 UK Census Data*, Working Paper 95/5, School of Geography, University of Leeds, Leeds.

Booth, C. (1887) Condition and occupation of the people of East London, *Journal of the Royal Statistical Society*, **50**, 326–401.

Booth, C. (1888) Condition and occupation of the people of East London and Hackney, *Journal of the Royal Statistical Society*, **51**, 276–339.

Booth, C. (1902–3) *Life and Labour of the People of London*, Macmillan, London.

Bracken, I. and Martin, D. (1989) The generation of spatial population distributions from census centroid data, *Environment and Planning A*, **21**, 537–43.

Brown, P. (1991) Exploring geodemographics. In *Handling Geographical Information* (eds, Masser, I. and Blakemore, M.) Longman, London, pp. 221–58.

Brown, P. and Batey, P. (1994) *SuperProfile Technical Note 2: Characteristics of SuperProfile Lifestyles and Target Markets: Index Tables, Pen Pictures and Geographical Distributions*, University of Liverpool, The Urban Research and Policy Evaluation Regional Research Laboratory, Liverpool.

Burrough, P. and McDonnell, R. (1998) *Principles of Geographical Information Systems*, Oxford University Press, Oxford.

Bush, G. (2003) Executive Order 13286, *Federal Register*, **68**, 10619–33.

Cameron, F. (2004) Techniques for Neighbourhood Classification, *New Representations: the use of geodemographic classifications in research and public service delivery conference*, Centre for Advanced Spatial Analysis, London. www.casa.ucl.ac.uk/lectures/ResearchMethods/downloads.html.

Carter, H. (1995) *The Study of Urban Geography*, Arnold, London.

Cathelat, B. (1990) *Socio-Styles: the New Lifestyles Classification System for Identifying and Targeting Consumers and Markets*, Kogan Page, London.

Chang, K.-T. (2003) *Introduction to Geographic Information Systems*, McGraw-Hill, New York.

Charlton, M., Openshaw, S. and Wymer, C. (1985) Some new classifications of census enumeration districts in Britain: a poor man's ACORN, *Journal of Economic and Social Measurement*, **13**, 69–96.

Cliff, A. and Ord, J. (1973) *Spatial Autocorrelation*, Pion, London.

Clinton, W. (1994) Executive Order 12906, *Federal Register*, **59**, 17671–4.

Curry, D. (1993) *The New Marketing Research Systems: How to Use Strategic Database Information for Better Marketing Decisions*, John Wiley & Sons, Inc, New York.

Curry, M. (1998) *Digital Places: Living with Geographic Information Technologies*, Routledge, London.

Database Marketing (2003) Monthly register rolls ahead, *Database Marketing*, **56**, 5.

Department for the Environment Transport and the Regions (1998) *Index of Local Deprivation*, DETR, London.

Department for the Environment Transport and the Regions (2000) *Regeneration Research Summary 31: Indices of Deprivation 2000*, DETR, London.

Department of the Environment (1987) *Handling Geographic Information: Report of the Committee of Enquiry Chaired by Lord Chorley*, HMSO Books, London.

Department of the Environment (1994) *A 1991 Index of Local Conditions*, DoE, London.

Diggle, P. (2003) *Statistical Analysis of Spatial Point Patterns*, Arnold, London.

Donnay, J.-P. and Unwin, D. (2001) Modelling geographical distributions in urban areas. In *Remote Sensing and Urban Analysis* (eds, Donnay, J.-P., Barnsley, M. and Longley, P.), Taylor & Francis, London, pp. 205–24.

Dugmore, K. and Moy, C. (eds) (2004) *2001 Census: Essential Information for Gaining Business Advantage*, The Stationery Office, London.

Farr, M. and Singleton, A. (2004) Applying Geodemographics to Education, *New Representations: the Use of Geodemographic Classifications in Research and Public Service Delivery Conference*, Centre for Advanced Spatial Analysis, London. www.casa.ucl.ac.uk/lectures/ResearchMethods/downloads.html.

Farrington, D. (2002) Developmental criminology and risk-focussed prevention. In *The Oxford Handbook of Criminology* (eds, Maguire, M., Morgan, R. and Reiner, R.), Oxford University Press, Oxford, Chapter 19.

Federal Geographic Data Committee (1997) *A Strategy for the NSDI*, FGDC, Reston, VA.

Feng, Z. and Flowerdew, R. (1998) Fuzzy geodemographics: a contribution from fuzzy clustering methods. In *Innovations in GIS 5* (ed., Carver, S.), Taylor & Francis, London, pp. 119–27.

Fischer, M. and Getis, A. (1999) New advances in spatial interaction theory, *Papers in Regional Science*, **78**, 117–18.

Flake, G. (1998) *The Computational Beauty of Nature: Computer Explorations of Fractals, Chaos, Complex Systems, and Adaption*, MIT Press, Cambridge, MA.

Flowerdew, R. (1991) Classified residential area profiles and beyond. *Research Report 18*, Lancaster University, North West Regional Research Laboratory.

Flowerdew, R. and Leventhal, B. (1998) Under the microscope, *New Perspectives*, **18**, 36–8.

Fotheringham, A., Brunsdon, C. and Charlton, M. (2000) *Quantitative Geography: Perspectives on Spatial Data Analysis*, Sage, London.

Fotheringham, A., Brunsdon, C. and Charlton, M. (2002) *Geographically Weighted Regression: the Analysis of Spatially Varying Relationships*, John Wiley & Sons, Ltd, London.

References

Fotheringham, A. and O'Kelly, M. (1989) *Spatial Interaction Models: Formulation and Applications*, Kluwer, Dordrecht.

Gittus, E. (1964) The structure of urban areas: a new approach, *Town Planning Review*, **35**, 5–20.

Goldstein, H. (2003) *Multilevel Statistical Models*, Hodder Arnold, London.

Gordon, D. and Forrest, R. (1995) *People and Places 2: Social and Economic Distinctions in England*, University of Bristol, Bristol.

Gordon, D. and Pantazis, C. (1997a) Appendix I: technical appendix. In *Breadline Britain in the 1990s* (eds, Gordon, D. and Pantazis, C.), Ashgate, Aldershot, pp. 269–72.

Gordon, D. and Pantazis, C. (1997b) Measuring poverty: Breadline Britain in the 1990s. In *Breadline Britain in the 1990s* (eds, Gordon, D. and Pantazis, C.), Ashgate, Aldershot, pp. 5–47.

Goss, J. (1995) Marketing the new marketing: the strategic discourse of geodemographic information systems. In *Ground Truth: the Social Implications of Geographic Information Systems* (ed., Pickles, J.), The Guilford Press, New York, pp. 130–70.

Hagel, J. and Singer, M. (1999) *Net Worth: Shaping Markets when Customers Make the Rules*, Harvard Business Press, Cambridge, MA.

Haggett, P. (2001) *Geography: a Global Synthesis*, Prentice-Hall, Harlow.

Hall, P. and Pfeiffer, U. (2000) *Urban Future 21*, E & FN Spon, London.

Hammond, R. and McCullagh, P. (1978) *Quantitative Techniques in Geography: an Introduction*, Oxford University Press, Oxford.

Harley, J. (1992) Deconstructing the map. In *Writing Worlds: Discourse, Text and Metaphor in the Representation of Landscape* (eds, Barnes, T. and Duncan, J.), Routledge, London, pp. 231–47.

Harris, R. (1999) Geodemographics and geolifestyles: a comparative review, *Journal of Targeting, Measurement and Analysis for Marketing*, **8**, 164–78.

Harris, R. (2003) Population mapping by geodemographics and digital imagery. In *Remotely Sensed Cities* (ed., Mesev, V.), Taylor & Francis, London, pp. 223–41.

Harris, R. (2003) An introduction to mapping the 2001 Census of England and Wales, *Society of Cartographers Bulletin*, **37**, 39–42.

Harris, R. (2004) Looking for (knitting) patterns: neighbourhood analysis and behavioural geographies in Bristol, England, *Applied Population and Policy*, **1**, 133–42.

Harris, R. and Chen, Z. (2004) Giving dimension to point locations: urban density profiling using population surface models, *Computers, Environment and Urban Systems*, in press.

Harris, R. and Johnston, R. (2003) Spatial scale and neighbourhood regeneration in England: a case study of Avon, *Environment and Planning C: Government and Policy*, **21**, 651–62.

Harris, R. and Longley, P. A. (2002) Creating small area measures of urban deprivation, *Environment and Planning A*, **34**, 1073–93.

Harris, R. and Longley, P. (2004) Targeting clusters of deprivation within cities. In *Applied GIS and Spatial Analysis* (eds, Stillwell, J. and Clarke, G.) John Wiley & Sons, Ltd, Chichester, pp. 89–110.

HEFCE (1998) *Consultation 98/39: Widening Participation in Higher Education: Funding Proposals*, Higher Education Funding Council for England, Bristol/London.

HEFCE (1999) *Report 99/24: Widening Participation in Higher Education: Funding Decisions*, Higher Education Funding Council for England, Bristol/London.

Howard, N. (1969) Least squares classification and principal components analysis: a comparison. In *Quantitative Ecological Analysis in the Social Sciences* (eds, Dogan, M. and Rokkan, S.), MIT Press, Cambridge, MA, pp. 397–412.

Howick, R. (2004) Building Neighbourhood Classifications. *New Representations: the Use of Geodemographic Classifications in Research and Public Service Delivery Conference*, Centre for Advanced Spatial Analysis, London. www.casa.ucl.ac.uk/lectures/ResearchMethods/downloads.html.

Hox, J. (2002) *Multilevel Analysis: Techniques and Applications*, Lawrence Erlbaum, Mahwah, NJ.

Hudson-Smith, A. and Evans, S. (2003) Virtual cities: from CAD to 3-D GIS. In *Advanced Spatial Analysis: the CASA Book of GIS* (eds, Longley, P. and Batty, M.) ESRI Press, Redlands, CA, pp. 41–60.

Hyndman, H. (1911) *The Record of an Adventurous Life*, Macmillan, New York.

Iliffe, J. (1999) *Datums and Map Projections for Remote Sensing GIS and Surveying*, Whittles Publishing, Caithness, Scotland.

Jackson, P. (1989) *Maps of Meaning*, Routledge, London.

Kearns, A. and Parkinson, M. (2001) The significance of neighbourhoods, *Urban Studies*, **12**, 2103–10.

Korner Committee, The (1984) *Fourth Report of the Steering Committee on Health Services Information*, Department of Health and Social Security, London.

Lawson, J. (2004) Think global – act local, *Database Marketing*, **61**, 12–16.

Lee, P., Murie, A. and Gordon, D. (1995) *Area Measures of Deprivation: A Study of Current and Best Practice in the Identification of Poor Areas in Great Britain*, Centre for Urban and Regional Studies, University of Birmingham, Birmingham.

Leventhal, B. (1993) Birds of a feather? Or, geodemographics – an endangered species? *Proceedings of the Market Research Society Conference*, pp. 223–39.

Liverpool City Council (1969) *Social Malaise in Liverpool: Interim Report on Social Problems and their Distribution*, Liverpool.

Lloyd, D., Webber, R. and Longley, P. (2004) Surnames as a quantitative evidence resource for the social sciences. In *Proceedings of Geographical Information Systems Research UK Conference* (eds, School of Environmental Sciences), University of East Anglia, Norwich, pp. 254–6.

References

Longley, P. and Batty, M. (eds) (1996) *Spatial Analysis: Modelling in a GIS Environment*, GeoInformation International, Cambridge.

Longley, P. and Batty, M. (eds) (2003) *Advanced Spatial Analysis: the CASA Book of GIS*, ESRI Press, Redlands, CA.

Longley, P., Brooks, S., McDonnell, R. and Macmillan, B. (eds) (1998) *Geocomputation: a Primer*, John Wiley & Sons, Ltd, Chichester.

Longley, P., Goodchild, M., Maguire, D. and Rhind, D. (eds) (1999) *Geographical Information Systems: Principles, Techniques, Management, Applications*, John Wiley & Sons, Inc, New York.

Longley, P., Goodchild, M., Maguire, D. and Rhind, D. (2001) *Geographic Information Systems and Science*, John Wiley & Sons, Ltd, Chichester.

Longley, P., Goodchild, M., Maguire, D. and Rhind, D. (2005) *Geographic Information Systems and Science*, John Wiley & Sons, Ltd, Chichester.

Longley, P. and Harris, R. (1999) Towards a new digital data infrastructure for urban analysis and modelling, *Environment and Planning B: Planning and Design*, **26**, 855–78.

Longley, P. and Tobon, C. (2004) Spatial Dependence and Heterogeneity in Patterns of Hardship: An Intra-Urban Analysis, *Annals of the Association of American Geographers*, **94**, 503–19.

Macintyre, S. (1997) What are spatial effects and how can we measure them? In *Exploiting national survey and Census data: the role of locality and spatial effects (CCSR Occasional Paper 12)* (ed., Dale, A.), University of Manchester, Manchester, pp. 1–17.

Martin, D. (1992) Postcodes and the 1991 Census of Population: issues, problems and prospects, *Transactions of the Institute of British Geographers*, **17**, 350–7.

Martin, D. (1998a) 2001 Census output areas: from concept to prototype, *Population Trends*, **94**, 19–24.

Martin, D. (1998b) Automatic neighbourhood identification from population surfaces, *Computers, Environment and Urban Systems*, **22**, 107–20.

Martin, D. (2000) Towards the geographies of the 2001 UK Census of Population, *Transactions of the Institute of British Geographers*, **25**, 321–32.

Martin, D. and Higgs, G. (1997) Population georeferencing in England and Wales: basic spatial units reconsidered, *Environment and Planning A*, **29**, 333–47.

Martin, D. and Longley, P. (1995) Data sources and their geographical integration. In *GIS for Business and Service Planning* (eds, Longley, P. and Clarke, G.), GeoInformation International, Cambridge, pp. 15–32.

Martin, D., Tate, N. and Langford, M. (2000) Refining population surface models: experiments with Northern Ireland Census data, *Transactions in GIS*, **4**, 343–60.

Mayor of London (2002) *London Divided: Income Inequality and Poverty in the Capital*, Greater London Authority, London.

McLuhan, R. (2003) Mapping customers, *Database Marketing*, **56**, 21–4.

Miles, S., Anderson, A. and Meethan, K. (eds) (2001) *The Changing Consumer: Markets and Meanings (Studies in Consumption and Markets)*, Routledge, London.

Miller, H., Han, J. and Fraser, S. (eds) (2001) *Geographic Data Mining and Knowledge Discovery*, Taylor & Francis, London.

Monmonier, M. (1996) *How to Lie with Maps*, University of Chicago Press, Chicago.

Monmonier, M. (2002) *Spying with Maps: Surveillance Technologies and the Future of Privacy*, University of Chicago Press, Chicago.

Morphet, C. (1993) The mapping of small area census data: a consideration of the role of enumeration district boundaries, *Environment and Planning A*, **27**, 267–78.

National Research Council (1993) *Towards a Coordinated Spatial Data Infrastructure for the Nation*, National Academy Press, Washington.

National Research Council (1994) *Promoting the National Spatial Data Infrastructure Through Partnerships*, National Academy Press, Washington.

National Research Council (1995) *A Data Foundation for the National Spatial Data Infrastructure*, National Academy Press, Washington.

Norman, P. (1969) Third survey of London life and labour: a new typology of London districts. In *Quantitative Ecological Analysis in the Social Sciences* (eds, Dogan, M. and Rokken, S.), MIT Press, Cambridge, MA, pp. 371–96.

ODPM (2004) *The English Indices of Deprivation 2004*, Office of the Deputy Prime Minister, London.

Office for National Statistics (2003) *Census Strategic Development Review. Alternatives to a Census: Review of International Approaches*, ONS, London.

Openshaw, S. (1984) *The Modifiable Areal Unit problem (CATMOG 38)*, Geo-abstracts, Norwich.

Openshaw, S. (1989) Making geodemographics more sophisticated, *Journal of the Market Research Society*, **31**, 111–31.

Openshaw, S. (1996) Developing GIS-relevant zone-based spatial analysis methods. In *Spatial Analysis: Modelling in a GIS Environment* (eds, Longley, P. and Batty, M.), GeoInformation International, Cambridge, pp. 55–73.

Openshaw, S. (1997) The truth about ground truth, *Transactions in GIS*, **2**, 7–24.

Openshaw, S. (1998) Building automated geographical analysis and explanation machines. In *Geocomputation: a Primer* (eds, Longley, P., Brooks, S., McDonnell, R. and Macmillan, B.), John Wiley & Sons, Ltd, Chichester, pp. 95–115.

Openshaw, S. and Abrahart, R. (eds) (2000) *Geocomputation*, Taylor & Francis, London.

Openshaw, S., Blake, M. and Wymer, C. (1995) Using neurocomputing methods to classify Britain's residential areas. In *Innovations in GIS 2* (ed., Fisher, P.), Taylor & Francis, London, pp. 97–111.

Openshaw, S., Charlton, M., Craft, A. and Birth, J. (1988) Investigation of leukaemia clusters by the use of a Geographical Analysis Machine, *Lancet*, **I**, 272–3.

Openshaw, S., Charlton, M., Wymer, C. and Craft, A. (1987) A mark I geographical analysis machine for the automated analysis of point datasets, *International Journal of Geographical Information Systems*, **1**, 335–58.

References

Openshaw, S. and Taylor, P. (1979) A million or so correlation coefficients: three experiments on the modifiable areal unit problem. In *Statistical Methods in the Spatial Sciences* (ed., Wrigley, N.), Pion, London, pp. 127–44.

Openshaw, S. and Turton, I. (1996) A parallel Kohonen algorithm for the classification of large spatial data sets, *Computers and Geosciences*, **22**, 1019–26.

Openshaw, S. and Wymer, C. (1995) Classifying and regionalizing census data. In *Census Users' Handbook* (ed., Openshaw, S.), GeoInformation International, Cambridge, pp. 239–69.

Orford, S., Dorling, D., Mitchell, R., Shaw, M. and Davey Smith, G. (2002) Life and death of the people of London: a historical GIS of Charles Booth's inquiry, *Health and Place*, **8**, 25–35.

O'Sullivan, D. and Unwin, D. (2003) *Geographic Information Analysis*, John Wiley & Sons, Inc, New York.

Park, R., Burgess, E. and McKenzie, R. (1925) *The City: Suggestions for Investigation of Human Behavior in the Urban Environment*, University of Chicago Press, Chicago.

Penny, N. and Broom, D. (1988) The Tesco approach to store location. In *Store Choice, Store Location and Market Analysis* (ed., Wrigley, N.), Routledge, London, pp. 106–20.

Pfautz, H. (1967) Sociologist of the city. In *On the City: Physical Pattern and Social Structure (Selected Writings of Charles Booth)* (ed., Pfautz, H.), University of Chicago Press, Chicago, pp. 3–170.

Pickles, J. (1992) Texts, hermeneutics and propaganda maps. In *Writing Worlds: Discourse, Text and Metaphor in the Representation of Landscape* (eds, Barnes, T. and Duncan, J.), Routledge, London, pp. 193–230.

Pile, S. and Thrift, N. (eds) (2000) *City A-Z*, Routledge, London.

Policy Action Team 18, The (2000) *PAT 18: Better Information*, The Stationery Office, London.

Raper, J. (2000) *Multidimensional Geographic Information Science*, Taylor & Francis, London.

Raper, J., Rhind, D. and Shepherd, J. (1992) *Postcodes: the New Geography*, Longman, Harlow.

Ravilious, K. (2004) Destined for destruction/welcome to fractal city, *New Scientist*, **2429**, 42–5.

Robinson, G. (1998) *Methods and Techniques in Human Geography*, John Wiley & Sons, Ltd, Chichester.

Robson, B. (1969) *Urban Analysis: a Study of City Structure*, Cambridge University Press, Cambridge.

Rogerson, P. (2001) *Statistical Methods for Geography*, Sage, London.

Sen, A. and Smith, T. (1995) *Gravity Models of Spatial Interaction*, Springer, Berlin.

Shevky, E. and Bell, W. (1955) *Social Area Analysis: Theory, Illustrative Application and Computational Procedure*, Stanford University Press, Stanford.

References

Shevky, E. and Williams, M. (1949) *The Social Areas of Los Angeles: Analysis and Typology*, University of Los Angeles Press, Los Angeles.

Simey, T. and Simey, M. (1960) *Charles Booth. Social Scientist*, Oxford University Press, London.

Sleight, P. (1997) *Targeting Customers: How to Use Geodemographic and Lifestyle Data in Your Business*, NTC Publications, Henley-on-Thames.

Sleight, P. (2004a) New to the Neighbourhood, *Database Marketing*, **57**, 35–40.

Sleight, P. (2004b) *Targeting Customers: How to Use Geodemographic and Lifestyle Data in Your Business*, World Advertising Research Center, Henley-on-Thames.

Sleight, P., Smith, G. and Walker, J. (2002) Optimising retail networks; a case study of locating Camelot's lottery terminals, *Journal of Targeting, Measurement and Analysis for Marketing*, **10**, 353–65.

Snijders, T. and Bosker, R. (1999) *Multilevel Analysis: an Introduction to Basic and Advanced Multilevel Modeling*, Sage, London.

Soja, E. (2000) *Postmetropolis: Critical Studies of Cities and Regions*, Blackwell, Oxford.

Theodorsen, G. (ed.) (1961) *Studies in Human Ecology*, Harper & Row, New York.

Tobler, W. (1970) A computer movie, *Economic Geography*, **46**, 234–40.

Treasury Committee (2002) *The 2001 Census in England and Wales*, House of Commons, London.

Unwin, D. and Fisher, P. (eds) (2001) *Virtual Reality in Geography*, Taylor & Francis, London.

Unwin, D. and Hearnshaw, H. (eds) (1994) *Visualization in Geographical Information Systems*, John Wiley & Sons, Ltd, Chichester.

Urban Task Force, The (1999) *Towards an Urban Renaissance: Final Report of the Urban Task Force*, E & FN Spon, London.

Voas, D. and Williamson, P. (2001) The diversity of diversity: a critique of geo-demographic classification. *Area*, **33**, 63–76.

Webber, R. (1975) *Liverpool Social Area Study, 1971 data: PRAG Technical Paper 14*, Centre for Environmental Studies, London.

Webber, R. (1977) *Technical Paper 23: an Introduction to the National Classification of Wards and Parishes*, Centre for Environmental Studies, London.

Webber, R. (1978) *Parliamentary Constituencies: a Socio-economic Classification: OPCS Occasional Paper 13*, OPCS, London.

Webber, R. (1985) The use of census-derived classifications in the marketing of consumer products in the United Kingdom, *Journal of Economic and Social Measurement*, **13**, 113–24.

Webber, R. and Craig, J. (1978) *Studies in Medical and Population Subjects, 35: Socio-economic Classifications of Local Authority Areas*, OPCS, London.

Webber, R. and Williamson, T. (2004) *Analysis for Youth Justice Board of Young Offenders' Offences in the County of Nottinghamshire*, available from Centre for Advanced Spatial Analysis, UCL. www.casa.ucl.ac.uk/lectures/ResearchMethods/downloads.html.

References

Weiss, M. (2000) *The Clustered World*, Little, Brown, New York.

Whitehead, J. (1999) Editorial, *Journal of Targeting, Measurement and Analysis for Marketing*, **8**, 108–10.

Wilson, A. (1974) *Models for Urban and Regional Planning*, John Wiley & Sons, Ltd, Chichester.

Wise, S. (2002) *GIS Basics*, Taylor & Francis, London.

Index

ACORN, *see* Classifications ACORN
Acxiom, *see* Vendors & suppliers
Advertising 4–5, 66, 69, 189, 205, 232
Aggregation 49, 90–2, 106–7, 115,
  145, 211–12, 215, 217–18, 222–3,
  252–5, 271
Atlantic City 5, 8–10, 12, 31
Atomistic fallacy 16, 245
Attribute information 16, 79, 81–2, 85,
  87, 89, 90, 92, 102, 106–7, 113, 115,
  221, 248, 252, 257, 271–2
  interpolating 70, 84, 112, 143, 204,
    247–8, 251–8
Australia 72, 148–50, 172, 187–8, 194,
  202, 205
Austria 191
Autocorrelation (spatial) 16–17,
  256–7, 263

Belgium 187, 191, 202, 205
Booth, Charles 24, 30–7, 45–51, 194
  Descriptive Map of London
    Poverty 30, 46
Bottom-up analysis 212, 249, 271
Brazil 149, 192, 197, 201, 205
Breadline Britain index 44
British Crime Survey 23, 141–2
Burgess model 37–8, 152, 200
Business models, practices & relationships
  54, 74, 235

CACI, *see* Vendors & suppliers
CAMEO, *see* Classifications ACORN
Canada 3, 67, 79–81, 85–6, 95–7, 101,
  113, 186–7, 190, 193, 200, 205

Catchment analysis 4, 10, 12, 19–22,
  55, 57, 84–5, 110, 116–19,
  133–8, 144–5, 198, 226,
  232, 247
Census
  Canadian 79–81, 85, 101
  comparisons of 189–95, 205
  data as input for geodemographic
    classification 3, 11, 17, 21,
    26, 35, 40, 45, 49–50, 55,
    59–61, 63–6, 73, 75, 90,
    148–60, 214, 234–5, 239,
    244–5, 273
  geographies 56, 69, 210–13, 262
  rolling 190–1
  UK 12, 35, 85, 100–1, 190–1,
    212–13, 240
  US 12, 39, 67–9, 85, 100–1
Census and Geodemographics
  Group 27, 236
Census bureaus
  Australian Bureau of Statistics 85
  Statistics Canada 85
  United Kingdom 23–4, 59–60, 62,
    65–6, 72, 191
  United States 61, 85
Central Postcode Directory 22, 56
Centre for Environmental Studies 22, 40,
  55, 75
Centroid 56–7, 165
Chicago School 24, 37, 50–1
China 187–8, 190, 192, 194–5, 197, 203–5
Chorley Report 58–9
Choropleth maps 90, 95, 98, 133
Claritas, *see* Vendors & suppliers

Geodemographics, GIS and Neighbourhood Targeting   Richard Harris, Peter Sleight and Richard Webber
© 2005 John Wiley & Sons, Ltd   ISBNs: 0-470-86413-3 (HB); 0-470-86414-1 (PB)